COM Programming by Example

Using MFC, ActiveX, ATL, ADO, and COM+

John E. Swanke

CMP Books
Lawrence, Kansas 66046

CMP Books
CMP Media, Inc.
1601 W. 23rd Street, Suite 200
Lawrence, KS 66046
USA
www.cmpbooks.com

Acquisitions Editor:	Berney Williams
Technical Reviewer:	Mike Wallace
Editor:	Michelle Dowdy
Layout Production:	Kristi McAlister and Kristy Peaslee
Cover Art Design:	Robert Ward

Distributed in the U.S. and Canada by:
Publishers Group West
1700 Fourth Street
Berkeley, CA 94710
1-800-788-3123
www.pgw.com

ISBN: 1-929629-03-6

VH #2 1-01

R&D Developer Series

For My Wife, Cathy Krinitsky,
My Great Woman

And for Paul Swanke,
A Friend Indeed

Acknowledgments

I'd like to thank everyone who worked on this book, directly or indirectly: Berney Williams for believing in this format, Paul Temme for his marketing acumen, and Michelle Dowdy and Kristi McAlister for making me sound like Hemingway (if he was a geek) and pulling it all together. I'd also like to thank Mike Wallace for making sure this book is more fact then fiction — and of course for keeping the manuscript out of Morley's hands.

Learning a subject is never done in a vacuum. I'd like to thank the people I learned COM with: my brother Paul, John Creaser, and Bob Otterberg,

I'd also like to acknowledge the entire gang at NeuVis, Inc. for creating an enterprise that's doing some pretty remarkable things with COM. NeuVis includes Rajiv Uppal, Jim Farrell, Devang Parikh, Jose Bescheider, Ritu Gupta, Tony "EJB" Tao, everybody who would be disappointed not to be mentioned here, and the genius behind NeuVis, Arun Gupta.

Table of Contents

Introduction

Welcome to the world of the Component Object Model (COM). If there was ever a subject that lent itself nicely to examples, it's this one. None of the API's, configuration files, classes, wizards, or helper applications of COM can stand alone — you need an example of how to use them all together in order to know how to do anything. And when you have an example and know what you're doing, adding COM to your application will take up almost none of your development time.

With this book, I have tried to provide examples of the most commonly used features of COM. I hope this format will be both informative to a COM novice yet still invaluable to a proficient COM expert.

What's In Store

The examples in this book are grouped into chapters that cover several different aspects of using COM, from in-process DLL's to remotely accessed applications, and from using the COM API directly (macho COM) to letting the Active Template Libraries (ATL) classes do most of the work. Similar chapters have been organized into one of the following sections.

Section I: COM Basics

Although I've tried to make this book example oriented, knowing a little bit of the story first can really help out when something you try doesn't work.

You can certainly skip this section if you want, but I will be referring back to it later.

Section II: COM Examples

The remainder of this book is composed of working COM examples representing the earliest days of COM to the present, from ActiveX Controls to MTS and COM+. Topics include creating and accessing COM objects using the API, MFC (Microsoft Foundation Classes), ATL, Visual Basic, or Visual J++ with considerations for multitasking, inheritance, and callbacks, among others.

About the CD

Included with this book is a CD containing a working Visual C++, Visual Basic, or Visual J++ v6.0 project for every example in this book. If you want to find the project for an example, just locate its example number among the subdirectories on the CD.

About the SampleWizard

Also on the CD is the SampleWizard utility, which can help you add the examples in this book directly to your applications. This utility guides you through a catalog of examples which if selected details the instructions and code necessary for including the example in your project. You will also be given the opportunity to substitute the example's project name ("Wzd") with your own.

The SampleWizard can be found in the \SWD directory on the CD. It makes use of the \Wizard subdirectory found in each example on the CD and contains all of the particulars for that example. Simply execute SW.EXE. The rest should be intuitive.

I have found it particularly useful as a user tool in the Developer Studio. Make sure to configure the directory as the current project directory [$(WkspDir)] and your requested example will be copied directly into your project directory.

COM Basics

Using COM usually boils down to knowing how to do three things: create a COM object, talk to a COM object, and everything else that gets in your way trying to do the last two items. Fortunately the "everything else" is usually not a big problem but can include what to do when your COM objects are used in a multitasking application and how do you prevent just anybody from accessing your objects.

The chapters in this section include the following:

COM Objects

In Chapter 1, we take a look at how to use the COM API to create a COM object. We also look at just what's involved in preparing your DLL or EXE so that the COM API can access the class objects inside. We do this using macho COM (using the API directly), the COM classes of MFC, and ATL's classes, both of which wrap the COM API. We also look at the benefits of creating all your COM objects as DLL's and letting MTS/COM+ manage them for you.

COM Communication

Having created a COM object, we now exchange data with it in Chapter 2. We will find that data exchange can be no more complicated then passing

1

arguments in any function call. Or it can entail an automatic serialization and reconstruction of your data through a network port.

Other COM Issues

In Chapter 3, we will review all the other headaches that come about when you try to create and communicate with a COM object. Included here is how to derive the functionality of another COM object in your own, how do you keep the data in your application thread-safe when your COM objects are in different threads, and how do you keep your COM objects secure from intruders trying to access your database.

COM+

In this final basic chapter (4), we review the newest enhancements to COM. Contrary to its name, COM+ is not the next version of COM, but the next version of a certain portion of COM called the Microsoft Transaction Server (MTS). Together they represent the cutting edge application of COM to allow you to write applications that can be massively scaled, potentially allowing you to replace a mainframe with a room full of PC's — or at least that's the hope. What did they say about a room full of monkeys with type-writers?

Chapter 1

COM Objects

In this chapter, we will review the nature of a COM object — what it is, how to create one and access its functionality, and how to put your own functionality into one. We will examine the API that allows you to work with COM objects as well as the many ways you can add COM to your application.

What is COM?

Simply put, COM is a system API that allows your application to access the functions and data in another application (EXE) or a dynamically linked library (DLL). But wait a minute, you might say. You could already do that using windows messaging, pipes, sockets, dynamic data exchange (DDE), remote procedure calling (RPC), and a few other system API calls.

And you're right. COM doesn't offer your application anything new except for one thing: a client/server standard. Because unlike any protocols and functionality you could come up with on your own, when you put your application into a COM object, anyone else following the standard can access it. (Not to mention your own client/server is probably filled with bugs anyway.)

COM actually stands for Component Object Model as an attempt to create an analogy with the hardware inside your computer. The question is put like this: "Why can't software be more like hardware? If I need a bus controller chip to design my computer, I can go out and buy one. I don't need to design one from scratch. Why not assemble my applications the same way? If I need a spell checker, I'll go out and buy a spell checker component."

This analogy, of course, makes most chip designers chuckle. The last thing they want is for their chip to be freely interchangeable with somebody else's. So why would a software company want to make their software freely interchangeable with someone else's? In reality, you'll find that although COM makes interfacing with someone else's software easier, the connection is still very much proprietary.

Another thing you'll quickly find with COM is that marketing had a field day with the terminology. Rather then helping to promote understanding, you'll find in most cases it promotes mystery. Marshalling? Aggragates? Free-threaded? What's going on there? OLE, ActiveX, Remotable Objects — which one means what this week? And most discussions of COM start off by listing a few more new buzzwords, which are then used throughout. Imagine a surgeon entering an operating arena with a handout of new terms for everything in the room and then proceeding to operate with those terms.

Why Was COM Developed?

COM evolved out of the solution to a specific problem — how to allow a word processor to edit the spread sheet in a letter without the user having to exit the word processor to do it. This solution was called OLE (Object Linking and Embedding) and contained a very sophisticated API that allows you even today to edit a letter within Internet Explorer using Microsoft Word. Most of this functionality is found in the OLE32.DLL.

However, with the second release of OLE (OLE2), its creators modified OLE32.DLL to expose more of the basic functionality of OLE so that its API could be used to allow any application to access the functionality of another. This API, which hasn't changed much since 1993, is what is considered to be the COM API.

When exposing this API, the OLE team modeled a lot of their design on the open system standards of the time. The purpose of these standards was to allow data and functionality to be freely interchanged by systems created by different software vendors. You will find remnants of this standard in COM, but make no mistake about it, COM is a Microsoft solution. If the COM API doesn't exist on your UNIX system, it doesn't get to play.

What are the Other Uses of COM?

Once the COM API was exposed, Microsoft used it to create what were then called OLE Controls to allow sophisticated controls (e.g., buttons that spiral, list boxes that play music) written in Visual C++ to be accessible to Visual Basic applications. Eventually this same technology was used to allow a control to be downloaded and used by your web browser. At this point, the name changed to ActiveX Controls or just plain ActiveX purely for marketing reasons. Although you can package any functionality in an ActiveX Control (this is still the only way to create a COM object in VB), its requirements have been specialized for supporting a user interface control.

At the same time, Microsoft encouraged all C and C++ developers to use the straight COM API to provide the infrastructure for their own client/server applications. Rather than write their own, the API provided a clean, standard and somewhat transparent way to exchange data and functionality with a server.

Another lesser use was for C/C++ applications that used DLL's. If the DLL's were optionally pulled in by the application using the `LoadLibrary()` API call, COM offered a more robust approach. It also provided a way to update a site without having to send an entire application update. You could just send the enhanced DLL.

COM also allowed multilanguage development (Visual Basic and C — not French and Portuguese). You could divide a project up into component DLL's and everyone could choose for themselves what language to write their component in — although other language considerations have limited this approach in practice. For example, you don't want to write a math component in fat, sloppy Visual Basic.

All of these applications of COM are still valid today, but they are quickly being dwarfed by its applications in the world of Information Technology (IT), where you may never find an entirely self-contained application again. Instead, applications have been divided into multiple COM objects that can run anywhere on a network. We'll be looking into this application in "What is COM+?" (see page 86) which gets its very own chapter (Chapter 4).

How Does COM Work?

If you write in Visual Basic or Visual J++, you don't need to care how COM works — these languages do their best to hide what's going on so you don't have to worry about it. C++ programmers, on the other hand, aren't so fortunate. For the same reasons that C++ is a faster, more versatile language than the others, it also requires that you have a more intimate knowledge of what COM is doing — although it has gotten a lot better in recent years. So if you don't want anything to do with C++ and aren't curious about how COM dynamically instantiated classes, please feel free to skip down to those sections in this chapter.

The Future of C++

A lot of COM books today give short shrift to C++. Just the other day I heard it referred to as a lower level language. VB and VJ++ are much easier to program in, creating more robust applications faster. But, as mentioned, they're also twice as slow and as one of my managers put it "like programming with a crayon". But aren't dumbed down languages in the future of most programmers? If you're operating a space ship in 2112 with gillions of lines of code can you afford to look for an exception error in a module way at the bottom? In fact, managers seem to be trading performance and flexibility for robustness and ease of use. As Stalin put it, "Quantity is a quality all its own." Look for applications of the future to have lots more functionality but with lots less features — if you know what I mean. (Windows Explorer can't even show you two subdirectories at the same time but will show that one directory on any device and without crashing.)

What follows then is a quick overview of what goes on when you create a COM object using C++ (and internally with other languages):

1. First, you call the `::CoCreateInstance()` COM API call, either directly or through a C++ class that wraps it. You specify two ID's, one for the class you want to instantiate and one for the DLL or EXE file it lives in. Very unique ID's are used instead of a class or filename to avoid the problem of having two COM objects with the same names on the same large system.

2. Since the DLL or EXE filename is specified with an ID, COM has to first convert it to an actual DLL or EXE filename by looking in the system registry where it was stored when the DLL/EXE was installed.

3. Now supplied with the actual filename, COM loads up the requested DLL using the `LoadLibrary()` API call or starts the desired EXE using the RPC (Remote Procedure Call) API.

4. Then COM tells the DLL or EXE which class to instantiate by using the other ID you specified to `::CoCreateInstance()`. Once the DLL or EXE has created an object, `::CoCreateInstance()` returns its pointer.

5. If the class lives in a DLL on your own system, calling its methods and passing arguments to it is not unlike any DLL method call using a class pointer, i.e., `p->Func(a,b);`.

6. If however you're calling a method on an object that was created in an EXE or on another machine, the arguments you pass to it go through the same serialization process that you would expect between any client and its server over a network — except COM does this for you automatically and almost transparently.

7. Destroying a COM object is a matter of telling the COM API you no longer need it by decrementing a reference count in the object. Once it reaches zero, COM destroys the object.

8. If all of the objects a DLL has supplied to your application are destroyed, COM unloads the DLL. If all of the objects an EXE has supplied any application on your system or the network are destroyed, COM stops the EXE.

Please see Figure 1.1 for this oversimplification of COM.

Figure 1.1   COM Overview

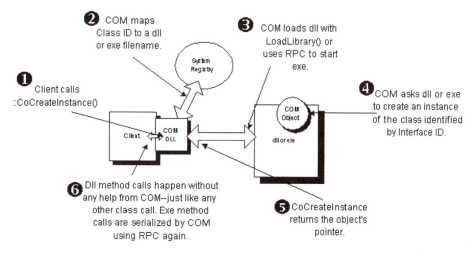

❷ COM maps Class ID to a dll or exe filename.

❸ COM loads dll with LoadLibrary() or uses RPC to start exe.

❶ Client calls ::CoCreateInstance()

❹ COM asks dll or exe to create an instance of the class identified by Interface ID.

❻ Dll method calls happen without any help from COM--just like any other class call. Exe method calls are serialized by COM using RPC again..

❺ CoCreateInstance returns the object's pointer.

❼ Client releases reference to object. COM actually destroys the object by unloading the dll or stopping the exe.

How Do You Create a COM Object Using C++?

Creating a COM object with C++ is actually easier than asking the system to create a window on your screen. Instead of specifying a litany of parameters like location, size, and procedure, you simply tell it what file the object is in and what object inside you want to create. The actual API call is as follows:

```
STDAPI CoCreateInstance(
REFCLSID clsid,              //Class id (what DLL or EXE file)
LPUNKNOWN pUnknown,          //Used for inheriting a COM object
DWORD dwContext,             //DLL, EXE local or remote
REFIID iid,                  //Interface id (what class in DLL or EXE)
LPVOID *ppv                  //returned pointer to created object
);
```

Rather then specify an actual class name or DLL/EXE filename, you use very unique id's called GUID's (Globally Universal ID), so unique that theoretically there's no two alike in the entire universe (please see the sidebar on GUIDs). Using a very unique id, instead of actual class and filenames, is to prevent two objects with the same name winding up on the same system and conflicting with each other.

Guids

A GUID is the Microsoft interpretation of the Universally Unique Identifier (UUID) proposed by the Open Software Foundation (OSF). A guid is a 128 bit number that is theoretically unique over time and space. To make it unique over space, it includes a fixed number that identifies the machine it was created on, usually a network card address. If there is no network card, another equally unique but constant number is located.

To make a guid unique over time, it includes a time stamp of when it was created. This stamp is the number of minutes that have elapsed since 1490. And just to be on the safe side, a guid also contains a random number generated just for the occasion.

You can create a guid using the GUIDGEN.EXE utility provided with MFC. The project and source files for GUIDGEN.EXE are also provided, but all this utility really does is make use of the COM API ::CoCreateGuid() to create guids.

Here then is the format of a GUID:

```
rrrrrrrr-tttt-tttt-oooo-aa-aa-aa-aa-aa-aa
```

where

rrrrrrrr	is a 32 bit random number
tttt-tttt	is the time stamp with the least significant 16 bit word appearing first
oooo	is related to the number of times the machine has been rebooted.
aa-aa-aa-aa-aa-aa	is a string of six bytes which are usually the address on the network card.

You'll find that guids come in three different and exciting formats. For use in the system registry,

```
{4323CD20-2559-11d2-9BD8-00AA003D8695}
```

To stuff a guid structure:

```
{ 0x4323cd20, 0x2559, 0x11d2, { 0x9b, 0xd8, 0x0, 0xaa, 0x0, 0x3d, 0x86, 0x95 } };
```

where the guid structure looks like this:

```
typedef struct _IID
    {
        unsigned long x;
        unsigned short s1;
        unsigned short s2;
        unsigned char c[8];
    } IID;
```

Or like this, to pass as eleven arguments to a function or macro:

```
0x4323cd20, 0x2559, 0x11d2, 0x9b, 0xd8, 0x0, 0xaa, 0x0, 0x3d, 0x86, 0x95
```

You'll notice that the degree to which the numbers of a guid change go from left to right with the numbers on the left changing the most to the numbers on the right not changing at all when generated on your machine. In other words, if you're looking to compare two guid numbers visually, look to the left.

If your application will require several guids, you may want to consider generating a range of them so that they appear consecutively in the system registry. You can do this automatically with:

```
guidgen /n50 /s >guids.txt
```

which would generate fifty (50) sequential guids, or you can do it manually just by incrementing the random number:

```
{4323CD20-...the same...
{4323CD21-...the same...
{4323CD22-...the same...
```

Unfortunately, the wizards get into the act of generating guids for you, so you may not have this option.

There are two id's used to create a COM object. The Class ID (CLSID) represents the DLL or EXE filename, and the Interface ID (IID) refers to the class within the DLL/EXE you want to create. Since these IDs are 128 bits long, you're actually passing arrays that you include as globals in your code.

CoCreateInstance converts the CLSID into an actual filename by looking in the system registry where these id's were associated with their filenames when the DLL or EXE was first installed. Most COM DLL's and EXE's have the smarts to do this themselves. Please see the sidebar on the other things you'll find in the system registry.

System Registry

COM information can be found in the system registry under the HKEY_CLASSES_ROOT key which itself points to the information under the HKEY_LOCAL_MACHINE\SOFTWARE\Classes\ key. The majority of the listing you'll find directly under this key are file extension associations (e.g., .txt is associated with notepad.exe) and COM Program ID's, but buried in this listing, you'll also find three other COM subkeys: AppID, CLSID, and Interface.

Program ID's

The COM Program ID's such as Access.Application or Access.Application.8 associate a long Class ID with this much shorter name for easier use in your application. In the case of Access.Application.8, Access represents the DLL or EXE file the class lives in, Application represents the class, and the ".8" specifies that this is the eighth version of this COM class. A Version Independent Program ID doesn't have a number at the end and is assumed to represent the latest and greatest class. Although a Program ID is much easier to use than a class ID, it also isn't as unique, and another COM class designated "Access.Application" installed on your machine will wipe out the registration of the current version. In the case of Visual Basic or VJ++, you have no choice but to use a Program ID. However, a program ID can be up to 39 characters long, so you can make it the same guid it represents. In other words, a program ID of "{98576D10-..." can represent a Class ID of {98576D10-... .

CLSID
Under this subkey, you'll find every Class ID registered on your machine in GUID form. Underneath each of these GUID keys, you'll find everything you need to know about that Class ID, the most important of which is the DLL or EXE filename associated with it. If it's a Class ID for a DLL, you'll find a InprocServer32 subkey that contains the DLL filename. For an EXE, you'll find either a LocalServer32 subkey with the name of an EXE file on your system or an AppID pointing to an entry in the AppID key for a remote EXE.

AppID
Under this subkey, you'll find an entry for every COM EXE or DLL you'll be running from another system. Under each AppID is a collection of the options you've configured for running this remote COM object, including where to find it and what the EXE or DLL filename is.

Interface
Under this subkey, you'll find the GUID's for all of the classes registered on your system. And under each of these GUID keys, you'll find the Class ID of the proxy/stub that would be used to pass arguments to and from that class when it's in another application. (See Chapter 2 for more about proxy/stubs.)

If the class lives in a DLL, CoCreateInstance() uses LoadLibrary() to load the DLL into your application. It then asks the DLL to create the class specified by the Interface ID and returns the pointer to you. You call methods using this pointer by first type casting it with a class definition of that class as seen here:

```
IMyComClass *p=(IMyComClass)::CoCreateInstance(…);
p->Method1(…);
```

We'll see how this class type is created below, but suffice it to say, you can now call any of the methods of this DLL as if it was a class that existed in your own application.

If the class lives in an EXE, CoCreateInstance() executes the application, then asks it to create the class specified by the Interface ID and returns the pointer to you. And you again call its methods as if the class existed in your own app.

For an example of using the COM API to create a COM object, please see Example 1 (page 104).

How Do You Communicate with Your COM Object Using C++?

That's right — calling the methods of a COM DLL or EXE is no different than calling the methods of any C++ class object — even if that EXE is on a system 500 miles away. But, wait a minute, you might say — sure, calling methods in a DLL is that simple — you've done it before. But when calling methods in another application, how does COM get around that process boundary and machine boundary that forces you to use RPC and your own client/server protocol just to get an elementary communication channel open — let alone make it a simple method call?

In fact, although COM does nothing extra for a COM DLL, a whole lot of behind the scenes processing is going on when you call the methods in a COM EXE server. In fact, if you bothered to use your debugger to step into a call to a COM EXE method, you'd find yourself suddenly up to your ankles in strange looking assembler instead of the method you're trying to call. That's because instead of calling the method you thought you were calling, you are in fact calling a go-between function called a *proxy*. This proxy is actually about to serialize your calling arguments into the standard COM client/server protocol and stick it out onto a channel to the COM EXE where another function called a *stub* unpacks it and finally places the call to the method. In other words, it's about to do everything you used to have to do manually to communicate with another application. For values returned by the method, the stub and proxy switch roles.

> **Note:** Actually, to step into the proxy, you need to do some fancy foot-work because the Visual Studio debugger is smart enough to skip over this code and take you directly to the server method. To see the proxy at work, you need to open up the disassembly view of your code and step through the assembler until _IT_ takes you into the proxy at the "call" opcode.

So where did these magical proxy and stub functions come from and how did you wind up calling them instead of the server method? COM pro-vides them automatically. When CoCreateInstance() created the object, not only did it start up the EXE and create the object, but rather than return its pointer back to you (which wouldn't do you any good anyway), it returned a pointer to the proxy function instead. So depending on what type of COM object you created, you could be getting a pointer to the real object or just a proxy function.

So where did this proxy function come from? From a proxy/stub DLL. CoCreateInstance() finds this DLL by mapping the Interface ID to a DLL filename under the Interface key in the System Registry. It then loads the same DLL into both your application and also the COM EXE server.

To see how this works, please see Figure 1.2.

Figure 1.2   Proxy/Stub Overview

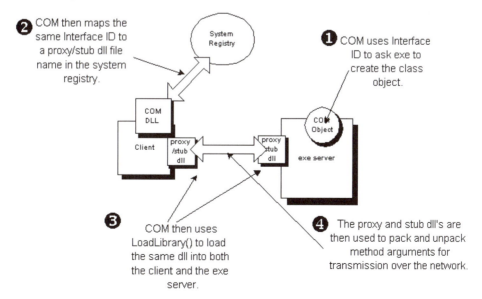

❷ COM then maps the same Interface ID to a proxy/stub dll file name in the system registry.

System Registry

❶ COM uses Interface ID to ask exe to create the class object.

COM DLL

Client

proxy /stub dll

proxy stub dll

COM Object

exe server

❸ COM then uses LoadLibrary() to load the same dll into both the client and the exe server.

❹ The proxy and stub dll's are then used to pack and unpack method arguments for transmission over the network.

So who wrote the proxy/stub DLL? You did — or at least you can at the same time you build your own COM object, or you can just use the standard proxy/stub that comes inside of `OLE32.DLL` — but not if your method requires something fancy be passed to it. There'll be much more on the proxy/stub DLL later in this chapter as well as in the next.

Cautions and Warnings

Even though this infrastructure does a pretty good job of hiding the fact that you're making an interprocess call, you really can't ignore what's going on. As an example, when you pass the pointer to a 20,000 byte array to a COM EXE method, you might think you're just passing four bytes. In fact, you're passing the whole 20,000 byte array — there and back! That's because that pointer doesn't mean anything in the other application, especially if the application is in Toledo. So the whole array has to be passed, referenced by that pointer there, and since any or all of it might change in the server method, the whole thing has to be serialized back to where it came from.

Another hazard in ignoring this infrastructure is passing uninitialized pointers. Since the proxy/stub function is a mindless robot, it's going to try and serialize anything that the meaningless pointer is pointing to. If it thinks the pointer is a NULL terminated string, it might try to serialize half of your memory to Toledo.

For more on why you can't exchange memory pointers between applications, please see the following sidebar on Process Boundaries.

Process Boundaries

The program address that you find in a data pointer (e.g., `char *p`) is not a pointer to an actual physical address in memory. Instead, it's a logical address that the CPU assigned your application when it brought the application into memory. If the CPU were to assign your application an actual physical address, it would be constantly modifying your application whenever it moved it in memory to make room for another application. As a result, the logical address you'll find in a pointer would be a worthless token to another application. Thus the need to bend over backwards to pass data between applications.

Some COM Terminology

In COM-speak, a COM DLL is considered to be "in-process" and a COM EXE is considered to be "out-of-process" for obvious reasons. A COM EXE executing on your system is considered to be a "local" server while a COM EXE you access on another system is considered to be a "remote" server. When using COM to access a COM EXE on another system, you are technically using DCOM (Distributed COM) instead of regular COM.

How Do You Destroy Your COM Object?

Destroying a COM object is a bit more complicated then destroying a normal C++ object. If an object was created out of a COM DLL for internal consumption by an application, it would be easy for COM to keep track of it and support a simple delete operative. However, a COM object might be created in a COM EXE on some other machine with its pointer conceivably shared among several applications.

So instead of adding all the additional overhead to keep track of that object in the OLE32.DLL, the object itself uses a reference counter scheme. Once created, the reference count starts at one. Any time you give that pointer to another function or application that might be using it for awhile, you need to increment it by one. So you don't actually delete an object, but you do decrement its reference count. COM then automatically destroys the object when the reference count reaches zero. COM also monitors whether all the objects created from a particular COM DLL or EXE server have been destroyed so that it can unload the DLL or terminate the EXE. Please see Figure 1.3.

Therefore, to "delete" a COM object you decrement its reference count by calling the object's Release() method:

```
p->Release();
```

And when passing a pointer to another function, you increment the reference count by calling:

```
p->AddRef();
```

Note: COM will automatically clean up your COM objects too (aka, "garbage collection"). In the case of a DLL, once your application terminates any object it created, it will be destroyed as with non-COM. And a COM EXE server continually checks to make sure that the client it created an object for is still there by sending it a message in the background. If the client fails to respond, the COM EXE server automatically decrements the object's reference count.

Figure 1.3 Deleting COM Objects

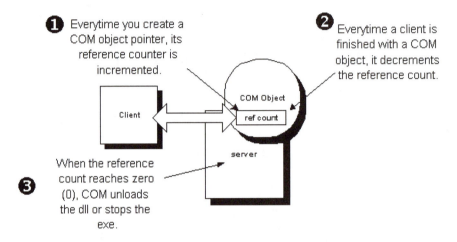

❶ Everytime you create a COM object pointer, its reference counter is incremented.

❷ Everytime a client is finished with a COM object, it decrements the reference count.

COM Object

ref count

Client

server

❸ When the reference count reaches zero (0), COM unloads the dll or stops the exe.

❹ COM also the clients of exe servers to see if they're still active and if not will automatically decrement the reference count.

How Do You Create a COM Object Using #import and Smart Pointers?

Creating and deleting a COM object got that much easier with the #import directive and smart pointers — new with version 6.0 of VC++.

The #import directive helps resolve the problem of keeping track of all those hard-to-manage GUID's. If you remember from above, the actual GUID's you need to identify the DLL, EXE, and class with are 128 bits long. What C++ programmers used to do was to include a .c file in their project that contained the numbers they needed as global arrays. (The .c file is generated when a COM object is built.) But now, with the #import directive, you can just import the COM object information directly into your code from a type library (also generated when a COM object is built) or the actual DLL or EXE file itself.

> **Note:** I may have said C++ programmers used to have to use .c files in the past, but considering the rapid public development of COM and the need to support legacy applications, all of the technologies mentioned in this book will probably be in use for years to come.

Smart pointers resolve the second problem of actually having to deal with an API call when you're programming in C++. An ordinary smart pointer just deletes whatever it points to when the smart pointer is destructed. The #import directive creates a custom smart pointer class for each of the Interface ID's it encounters while importing a type library, DLL, etc. that wraps the functionality of CoCreateInstance(), Release() and AddRef() too. This custom class already contains the Interface ID GUID so that all you have to provide when creating a COM object is a Class ID. As an example, if the class you wanted to create was called IComClass, the #import directive would create a smart pointer class called IComClassPtr. Then all you need to create a COM object from this class is the GUID of the DLL or EXE:

```
IComClassPtr cls(
            __uuidof(XXX) // the DLL or EXE GUID
            );
```

where the definition of *XXX* is added to your source code by the #import directive.

To access the COM object's methods using this class, you would use:

```
cls->Method();
```

Notice we're using pointer syntax here instead of cls.Method(). Smart pointers overload the pointer syntax to indicate you are accessing the method of the COM object it's wrapping as opposed to the smart pointer's own methods. But what if you wanted to create cls in global memory with:

```
IComClassPtr *pCls=new IComClassPtr(XXX);
```

You would then access the COM object's methods by first de-referencing the pointer with:

```
*(pCls)->Method();
```

Now you can destroy a COM object as simply as you can a C++ object — just by destroying the instance of the smart pointer:

```
delete pCls;
```

For an example of using #import and smart pointers to create a COM object, please see Example 2 (page 110).

How Do You Write a COM Object Using C++?

So far, we've only talked about creating an instance of a COM object written by someone else, so how do you write one yourself?

Writing a COM class in C++ starts out the same way any C++ class is written — by defining its member variables and functions. But instead of defining these members in a standard C++ .h file, you define them in an Interface Definition Language (IDL) file.

The IDL File

As you can see from the sample IDL file in Figure 1.4, some of the syntax you use is the same as in a standard .h file. However, additional syntax was needed to help define how you eventually want to create the object as well as how you would like the proxy/stub DLL to receive and send your calling parameters.

Figure 1.4   Sample IDL File

```
[
        object,
        uuid(E1637ED6-1746-11D2-9BC7-00AA003D8695)
]
interface IWzd : IUnknown
{
        HRESULT Initialize();
        HRESULT method1([in] short nIn,
                                [out] char *pOut,
                                [in, out] char *pInOut );
        HRESULT method2([in, out] char *pInOut);
};
```

> **Note:** Although the IDL file is a separate entity in V6.0 of VC++, V7.0 will allow you to define the additional IDL syntax right in a standard .h file. To differentiate between C++ syntax and IDL, all IDL syntax is bracketed with braces (i.e., []).

In C++ terms, what you are defining with your IDL file is a pure virtual C++ class with all public member functions and no member variables. And, in fact, when an IDL file is compiled, one of the files generated is a standard C++ .h file that contains this pure virtual class. You can then use this .h file directly to type cast the pointer returned by CoCreateInstance() as seen earlier. On the server side, this pure virtual class forces you to implement every method defined — otherwise the object wouldn't compile.

By convention, names for these pure virtual classes are prepended with the letter "I" and the classes themselves are called Interfaces instead of classes. But conceptually, it's probably easier to visualize them as just a pure virtual class that keeps the client and server synchronized.

IDL files are compiled using the Microsoft Interface Definition Language (MIDL) compiler, which you are no more likely to invoke directly than you would the C++ compiler. And instead of generating a binary object file, the compiler generates several binary and source files. An IDL file called MyInterface would cause MIDL to generate the following files:

MyInterface.h

This is the actual C++ abstract class definition. You derive your server object from this class and use it to typecast a COM object pointer in your client.

MyInterface_i.c

This file contains the GUID's that identify a DLL/EXE using this interface to your client. Since the GUID is actually an array of 128 bits, it must be included in your client as a global constant that's passed to the API as a pointer to this array — quite a mess to get such uniqueness and fortunately, unnecessary when you use #import and smart pointers.

MyInterface_p.c,dlldata.c

These additional files can be used to create a proxy/stub DLL for your COM object. This is the proxy/stub DLL mentioned earlier that helps your COM object perform interprocess communication and will be discussed in detail in the next chapter. Suffice it to say for now that you don't even need to create it unless your object needs to pass an argument that's not already in OLE32.DLL's repertoire.

MyInterface.tlb

A binary called a *type library* defines the information in MyInterface.h (after all, an .h file is only good to a compiler). When OLE32.DLL needs the definition of your COM class, it turns to this binary file.

There is no IDL syntax to add a member variable to your interface class. Because accessing the member variables of a COM object from a client suffer from the same process boundary problems that functions do, it was decided that the member variables of a COM object would be accessed using new set and get functions. That way only one interprocess communication infrastructure had to be developed. The member variables of a COM object are also referred to as the *properties* of that COM object.

IUnknown

Another aspect to writing a COM class is that unlike a standard C++ class is that all COM objects have the same base class for the usual C++ advantages (every COM object will provide the same limited function set and all COM objects can be used polymorphically). This base class is called IUnknown, the reasons for which are unknown. A more accurate name might be along the lines of MFC's CObject — perhaps IComObject.

We've already covered two of the functions that IUnknown defines: Release() and AddRef(). The third and last function is called QueryInterface() which allows you to ask an object if it also contains a COM class you specify with an Interface ID — if it does, you get a pointer to it, else you get an error back.

And notice I said that IUnknown *defines* these functions — it doesn't actually *implement* them. That's your job as the author of a COM object — but don't worry about it, you can get help doing it.

Binding

Another choice when writing a COM object is how your client will bind to your object. In other words, will the client have access to information on your object at compile time (in the form of IDL output files) or could the client have been written months before and therefore not had access.

When a client does have access to the IDL information, specifically the class type override, your compiler fills in the addresses of any methods you call in a technique called *binding* or *early binding*. But when this information isn't available, you can still build functionality into your COM object so that a suitably prepared client can actually query your object at runtime to find out what methods it supports and what argument those methods require in a technique called *late binding*.

To support late binding in your object, you have to implement yet another interface called `IDispatch` — and yet again you have help. You are also forced to use just the calling arguments that `OLE32.DLL` supports (you can't use your own proxy/stub). Most developers add support for both ways to bind to their object.

The plus side to late binding is that you have total separation between client and server — they don't know about each other until show time. On the other hand, calling a method is a lot slower because of this querying.

Late binding was originally called automation because it was used by one application to automatically control another application. As an example, a configuration application might tell a word processor to spell check a document for it.

Please see Figure 1.5 for a view of early vs. late binding. For an example of late binding, please see Examples 3 and 4 (beginning on page 116).

Figure 1.5 Early Binding vs. Late Binding

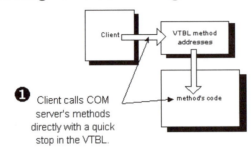

Early Binding

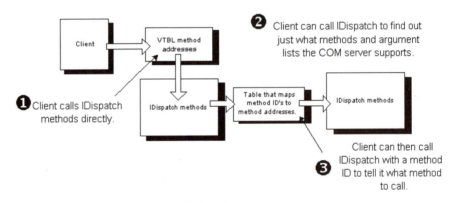

Late Binding

Other COM Object Types

Besides binding, there are several other choices when writing a COM class including: how many objects can be created from your class, where will it be created, and whether or not COM will make your object thread-safe.

Singletons

A COM class that returns a pointer to the exact same object no matter how many time you create and destroy it is called a *Singleton*. Singletons aren't used very often, but can be used to maintain global data among the COM objects of your COM server. COM has no built-in support for Singleton objects but you can find an example of one in Examples 15 (page 166) and 23 (page 213).

Single vs. Multiple Use

When your COM class lives in an EXE, you also have a choice of whether or not a new EXE has to be executed every time a new COM object is created. In other words, when a client uses `CoCreateInstance()` to create your class and your executable is already running, will COM start up a whole new EXE. A COM class that forces COM to start up a new EXE is called a *Single Use* class. Single Use classes are good in an SDI application where the application can only handle one document class at a time (if another document has to be edited, a new EXE has to be started). *Multiple Use* classes reside in MDI applications which can handle multiple instances of the document class. You specify single or multiple use in an API call your application makes when it first starts up. MFC makes this call automatically for you.

Running Object Table

Rather than create a brand new object, you can also allow clients to access an existing COM object by registering it in the Running Object Table (ROT) using a COM API call. A client can then use another COM API call to retrieve its pointer. This technique is mostly used in document applications where several applications can be working with the same document that's sitting in a COM object.

Thread Safety

You can have COM make your object thread-safe rather than build that support in yourself. Thread safety involves preventing two threads in your multitasking application from writing to the same data area at the same time. Before COM, you would have to make your application thread-safe by putting all accesses to the same data area in the same thread or by using mutexes that forced each thread to take a turn at the data. COM gives you the additional option of letting it force all objects that access the same data into the same thread — *no matter* what thread the object was created in. You choose this protection at development time by picking a threading model for your object that's stored in the system registry under its CLSID. There's much more on COM thread safety in Chapter 3.

DLL Host

And before we move on, don't think that if your server will be on another machine that you can't use DLL's for your COM objects. Yes, a DLL needs an executable to run, and a client can't directly load up a DLL that's sitting on another machine. But what if there was a friendly EXE sitting on that other machine that was willing to load up that DLL for your client and pass information between that object and your client. In fact COM provides both an API to create your own friendly DLL Surrogate application and it also provides one pre-written for you called DLLHOST.EXE.

This last technique of using a surrogate process to load DLL's on a remote machine may not look it, but it's the foundation of the strategy of using COM to revolutionize the way applications are written. Needless to say, there's lots more on this technique in Chapter 4.

Writing a COM Object in Straight C++

So now that we know just about all the options there are for writing a COM class, how do you write your own COM class in straight C++? The short answer is you don't — not in the real world, not with a schedule to keep. Even in the base case where you're just supplying your own functionality, you still need to implement a few classes (remember IUnknown and IDispatch from above) plus all the other overhead required by COM to load DLL's or execute EXE's. A COM DLL for example must provide a DllGetClassObject() and DllCanUnloadNow() function to allow COM to load and unload the DLL. Another class we haven't even talked about is IClassFactory which your DLL or EXE server must implement to actually create the COM object. In the case of an ActiveX server, there are dozens of classes and functions you need to implement in order to keep your client happy.

So instead of looking at how to write a COM object from scratch, we will be focusing on creating our COM classes with the assistance of two class libraries that Microsoft provides: the Microsoft Foundation Classes (MFC) and the Active Template Library (ATL) as well as their wizards.

MFC, in fact, has several classes that wrap the COM API making it easier to use. Unfortunately, MFC is more of a library of user interface classes (creating windows, etc.) and can make your object too bloated for an object that doesn't need this functionality, such as an object that just wants to access a database.

For those type of applications, there's the ATL — a class library that simply wraps the COM API without bringing along any extra baggage. You still need an ATL DLL, but this DLL's been built for speed and size.

And so we begin.

How Do You Write a COM Object Using MFC?

Creating an MFC COM object that just supports late binding is almost as easy to do as creating the same for a Visual Basic or VJ++ COM object. You add classes, methods, and properties without having to worry much about GUID's and IUnknown. However, by the same token, the resulting product is slower than having created your object using early binding and restricted to a certain set of argument types. On the other hand, adding early binding support to an MFC COM object can be a real chore. Let's look at the easy case first: late binding.

To create an MFC DLL to hold your COM class, you start by using the AppWizard to create a Regular DLL and pick the "Automation" option in Step 3. A Regular DLL is required because your COM object must be usable by non-MFC applications.

Creating an MFC EXE is even easier — just create your application as before, again picking "Automation" in step 3.

The AppWizard will then generate the ClassID for your DLL or EXE, generate your project classes and add the COM API calls to them for you. If you picked an SDI application, MFC will automatically register your COM classes as Single Use. For MDI, it will pick Multiple Use.

To add your COM class to either of these projects, you open the ClassWizard and create an MFC class derived from CCmdTarget. Once CCmdTarget is selected, those mysterious radio buttons at the bottom become enabled and you pick "Automation". To now add methods and properties (i.e., methods that access your COM class's member variables), you use the ClassWizard again, opening up the COM class you just created and selecting the "Automation" tab (notice a trend yet?). There you will find buttons that will allow you to add your classes methods and/or properties. You can then go to the class and add your functionality.

To register your COM DLL so that CoCreateInstance() can find it, you simply use the regsvr32.exe utility like so:

```
regsvr32 /s MyCOMServer.dll
```

All regsvr32 does actually is call a method that MFC added to your DLL that registers the DLL.

To register a COM EXE, you need to execute it. That's right — MFC stuck a COM registration routine into your application's InitInstance(). And having registered your COM object, your client is ready to use it.

Custom Interface

But what if you want to use early binding instead? What's the "real chore" mentioned earlier? You start by having to create your own interface project without the help of the AppWizard. (Actually, you can just use the project on the CD for Example 9). This project contains two source files — the IDL file you define your COM classes with and a make file that uses MIDL to compile it — and the C compiler and linker to create a proxy/stub DLL.

Once this Interface project is created, creating your DLL or EXE project is as easy as before. But this time when you use the ClassWizard to create your COM class, instead of specifying "Automation" at the bottom, you specify "Creatable by Type ID".

As with the Interface project, adding methods and properties to your COM class is also a manual process. You have to add several macros yourself as well as implement IUnknown (i.e., AddRef(), Release(), and Query-Interface()). But then again, hopefully Examples 10 and 11 (beginning on page 147) make it clear what to add.

Early binding in MFC is also known as adding a "Custom Interface". Late binding is also known as adding an "Automation Interface" or dispinterface. You can support both in the same COM class just by cross-referencing the method calls.

Don't worry — as we will see, adding early binding support with ATL is a lot easier. So why haven't they made it easier in MFC? Because early binding just doesn't make much sense for an MFC application. The MFC DLL is stuffed with user interface classes that are wasted space with the majority of number crunching COM objects. Sure, you also need COM objects that deal with the interface, but that's what ActiveX COM objects are for, and MFC does have an ActiveX Wizard that makes writing those COM classes easier. Let's just say that a lot of early COM programmers had to worry about early binding until ATL came along.

Note: There is one area where MFC early binding is still important — when it comes to COM EXE servers that you also want to support MFC. ATL allows MFC support in the DLL projects it creates, but you need to bend over backwards to get the same support in an EXE.

Please refer to Examples 10 and 11 (beginning on page 147) for details on creating a COM object using MFC.

How Do You Write a COM Object Using ATL?

Creating a COM object using ATL that supports both binding types is almost as easy as creating a VB object — if it wasn't for all those options.

To create either DLL or EXE, you start by invoking the ATL COM Wizard. You can add MFC support to a DLL project but not an EXE project (ATL's infrastructure for an EXE conflicts with MFC's). You can also combine the code necessary for creating a proxy/stub DLL right into your DLL or EXE.

One of the files generated for you by the ATL COM Wizard is an IDL file. But rather than edit this file directly to add your class definition(s), you use the Studio's "New ATL Object" command which presents you with several different types of COM classes to create — and a lot of options too. After a COM class is created, you can add methods and properties by going into the Class View tab of your Workspace View, locate the class you just added and right click on it. A popup menu gives you a choice of adding a method or property. Once entered, the Class View will automatically add the method/property to three files: the COM class's `.cpp` and `.h`, and the `.idl`. And if you make a mistake, you have to manually edit those same three files, because there is no "Undo".

Once you add your functionality to the methods, you're done. The project will automatically register this object when built, or you can use `regsvr32` again to register the DLL's or execute the EXE files to register those.

Please refer to Examples 18, 19, and 20 (beginning on page 192) for details on creating a COM object using ATL

So How Do You Create a COM Object with Visual Basic?

This will be short and sweet. To create a COM object using VB, you use the Projects/Components menu item to add the COM object to your VB project. If, as an example, the AppID for the DLL or EXE is MYDLL and the class name is `MyClass`, you would load the DLL/EXE and create the class simply by using the heavily overloaded Dim syntax:

```
Dim c As New MYDLLLib.MyClass
```

and use its methods with:

```
rval=c.Method(x,y,z)
```

To write a VB COM DLL or EXE, you just open a new project and pick ActiveX EXE, ActiveX DLL, or ActiveX Control and add your methods as public subroutines:

```
Public Sub MyMethod()
      :    :
   End Sub
```

No muss or fuss — but at a price of a performance hit. All of the background processing (e.g., VB only uses late binding) needed to automate this connection makes a VB COM object about 40% slower than an equivalent C++ object. On the other hand, if you have lots of VB programmers and fast platforms, this might be your solution. Or perhaps you can save C++ for just the CPU intensive objects and leave the rest for VB.

In a lot of COM applications, VB is used for the front end user interface while C++ is used for the actual COM objects. Another alternative to writing VB objects is to write objects in Java. Java also makes COM objects easy to develop — and with the same performance hit as VB — but the language is more flexible than VB and more robust then C++.

So How Do You Create a COM Object with Visual J++?

To create a COM object from VJ++, you start with the Project/Add COM Wrapper... menu item and select the existing COM class from the list — not unlike VB. Only this time a new Java class is created that takes care of all the minutia involved in creating and destroying the COM object — not unlike a smart pointer class.

To write your own J++ COM DLL (no EXE available), you just open a new project and pick COM DLL or Control. To define the methods and properties your COM class will support, you go into Project/Add Class... and pick Interface class. You define your COM class here — and to implement it — you just derive another class from it and fill it in with functionality. This time, J++ takes care of GUID's, IUnknown, etc.

For an example of using VJ++, please see Examples 29 and 30 (beginning on page 238).

What's the Service Control Manager?

In reading about COM, you'll occasionally see the Service Control Manager or SCM (pronounced "scum") mentioned. The SCM is located in RPCSS.EXE and just provides part of the functionality that's invoked whenever you call CoCreateInstance(). In particular, the SCM does the actual mapping of GUID to filename and in the case of EXE servers, it starts the EXE locally or remotely in cooperation with a remote SCM. The COM libraries located in OLE32.DLL do just about everything else.

Why the SCM is singled out in COM documentation is a mystery. I only bring it up in case you're curious. And if you're wondering if this is the same SCM that manages Services on Windows NT, the answer is no.

Summary

In this chapter we explored the creation and destruction of a COM object. We also looked at how hard it is to write one in C++ versus VB or VJ++; however, we also saw why C++ creates a faster COM object. We looked at the minutia involved with writing a C++ COM class — especially in the case of a custom MFC interface — but also saw how ATL takes care of a lot of the minutia for us.

One aspect to calling a COM method that I carefully avoided was how exactly COM automatically pass arguments to and from the method — especially when the COM object is on another machine, potentially in another country. We will see how it's done and what you need to make it work in the next chapter.

Chapter 2

COM Communication

In Chapter 1 we briefly reviewed how method arguments are passed to and from a COM object. We saw how COM automated interprocess communications using an IDL file we write to create a custom communications DLL. In this chapter we'll take a closer look at that IDL file with an emphasis on using it to optimize our communications with a C++ COM object. We'll also take a look what data types C++ programmers should be aware of when communicating with a Visual Basic or Visual J++ client. Since the IDL file is taken care of by VB and VJ++, there's no need to review communications between objects written in those languages — once again the solution is simple yet unchangeable.

How Do COM Objects Communicate?

As seen in the last chapter, after a COM object is created in an out-of-process DLL or EXE, interprocess communications with that object can go one of two ways. If all of the argument types are standard (as listed later in this chapter), the COM DLL itself (`OLE32.DLL`) will support communications with the object guided by the type library generated from your IDL file. If any of the arguments aren't standard, COM looks for and loads your own custom communications DLL that was built from the files generated once again by your IDL file.

In both cases when you make a call, your arguments are formatted into a standard communications protocol called Network Data Representation (NDR) transfer syntax. In fact if you look in the XXX_p.c file used to build your custom proxy/stub DLL, you'll find that 99% of it is your calling arguments formatted into line after line of NDR macros.

From there, COM uses the RPC API (RPCNSS.EXE) to call the OLE32.DLL or custom proxy/stub DLL on the server. The RPC API itself uses the network transport layer to actually communicate with the other DLL. That DLL unpacks your calling arguments and makes the actual call to the method. Returned arguments are repacked and resent back to the client. Please see Figure 2.1 for an overview of this process. Incidentally, sending and receiving data from a COM object is called "marshalling" the data.

Figure 2.1   Marshalling COM Data

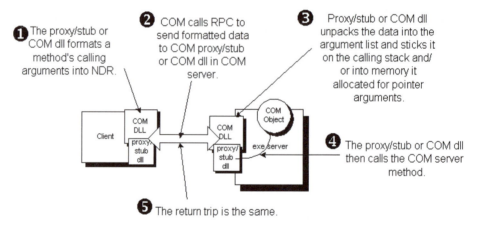

❶ The proxy/stub or COM dll formats a method's calling arguments into NDR.

❷ COM calls RPC to send formatted data to COM proxy/stub or COM dll in COM server.

❸ Proxy/stub or COM dll unpacks the data into the argument list and sticks it on the calling stack and/ or into memory it allocated for pointer arguments.

❹ The proxy/stub or COM dll then calls the COM server method.

❺ The return trip is the same.

As you can see in both cases, the IDL file plays a key role in how your calling arguments are passed to the object, but what exactly goes into an IDL file?

Basic IDL File Formats

There are three basic types of IDL files: one for a COM server that will be supporting an early binding, one for a COM server that will support late binding and one for a COM server that will support both.

Early-Binding Interface

In its most primitive form, an early-binding IDL file is almost indistinguishable from a C++ header file with the following differences:

- "Interface" is used instead of class.
- All COM classes (unlike C++) share the same base class IUnknown.
- The class, its methods and even its arguments are further defined with keywords that are specified within brackets (i.e., []).

These keywords or attributes are perhaps the most imposing aspect of an IDL file as seen here:

```
[
    object,
    uuid(E1637ED6-1746-11D2-9BC7-00AA003D8695)
]
interface IWzd : IUnknown
{
    HRESULT Initialize();
    HRESULT method1([in] short nIn,
                    [out] char *pOut,
                    [in, out] char *pInOut  );
    HRESULT method2([in, out] char *pInOut);
};
```

But remove all of the stuff between the braces ([]) and replace the IDL tokens with their C++ equivalents and you have this:

```
class CWzd : public CUnknown
{
    long Initialize();
    long method1(short nIn, char *pOut, char *pInOut  );
    long method2(char *pInOut);
};
```

The COM server created when implementing this interface is internally similar to any C++ class, allowing a client to directly call its member functions — for those times when it's in the same address space as the client.

Late-Binding Interface

When a COM server supports only a late-binding interface, it exposes only one class, IDispatch. The reason for this is that all clients in the future and in the past can always access a late-bound server because it will always have this same class interface.

IDL files used for a late-binding interface still create an address table of methods, but the client must use one of IDispatch's four methods to query and access those addresses by a simple ID you define in the IDL file.

There are three IDispatch methods used for querying an object for its methods and arguments and one for calling a method.

- GetTypeInfo() and GetTypeInfoCount() can be used by an inquisitive client to find out all of the methods and functions in a COM object that supports late binding. These methods allow a programmatic way for a client to look at an object's type library and return not only the ID of a method to call but also what should be in its argument list. But a COM server isn't required to return this information, which can be the case for a particularly sensitive application.

- GetIDsOfNames() can be used to match up a list of method names with their ID's, for a client that already knows the method name and argument list of its COM server but needs to know exactly which ID to call.

- And the Invoke() method is used to actually call the method when supplied with the method ID and the argument list in an array of DISPPARMS.

So unlike the IDL file in an early-binding interface, a late-binding interface must have at least one method attribute defining its id. Here it's shown as "id(1)" and "id(2)":

```
dispinterface CWzd
{
    properties:
        [id(1)] int property1;
    methods:
        [id(2)] HRESULT method1();
        };
}
```

The "properties:" seen above define methods that store and retrieve some "property" of a COM object. You can think of properties as the member variables of a COM class — although you're accessing them through a method, they don't necessarily represent a member variable. Which brings up the question: where's the member variable declarations for a COM class? There are none. Because a COM object doesn't necessarily reside with the client — it could, in fact, be in Chicago — you must always access its functionality through method calls, thus the invention of "Properties".

Dual Interface

As you might expect, a dual interface is an interface that can allow a COM server to support both early and late binding. The IDL file for a dual interface might look something like this:

```
[
        uuid(E1637ED6-1746-11D2-9BC7-00AA003D8695),
        oleautomation,
        dual
    ]
    interface IHello : IDispatch
    {
        [id(1)] HRESULT Initialize(); HRESULT method1([in] short nIn,
                [out] char *pOut, [in, out] char *pInOut  );
        [id(2)] HRESULT method2([in, out] char *pInOut);
    };
```

Notice that a dual interface contains the trappings of both interface types: an interface ID and "interface" designation of an early-binding interface and the method attribute id and IDispatch derivation of a late-bound interface.

And the methods that are exposed to a client are also a combination of both IDispatch's and your own, with IDispatch's methods at a fixed location followed by your own methods at any location.

The VTBL

As an application writer, you shouldn't care, but the locations I referred to above are in what's called a *virtual table* (VTBL) which is generated by the C++ compiler as an address table at the start of every class object and the members of that table point to every method in that object. With a COM server that supports both types of interfaces, the first three addresses of the VTBL will contain IUnknown's methods (remember every COM class is ultimately derived from IUnknown), followed by IDispatch's four methods, followed by all of your custom methods.

As I said, as an application writer you don't need to worry about this anymore than you do whenever you write a C++ application because when you create a COM server, you will normally be using MFC or ATL and these libraries take care of this table for you.

Type Library Declaration

At the bottom of an IDL file you might also find a Library declaration, which the MIDL uses to generate a Type Library. A client can use Type Libraries at compile time or runtime or anytime to determine what methods are supported by a COM class and what their argument lists contain. The IDispatch class's GetTypeInfo() method gets its information from this type library and C++'s "#import" directive uses this type library in lieu of the .h file that the MIDL compiler also generates. Even the OLE32.DLL uses this type library to find out what a method's argument list is so that it can transmit it between server and client.

The syntax of a type library declaration is similar to a class definition and listing all of the interfaces this COM class will be implementing:

```
library WzdTypeLib
{
    importlib("stdole32.tlb");
    importlib("stdole2.tlb");


    [
        uuid(DCBC68C9-4E2A-11D2-AB34-00C04FA3729B),
    ]
    coclass WzdClass
    {
            [default] interface IWzd;
    };
};
```

So why is it called a type library instead of an interface library or a class library? Because it's COM.

So far, we've only reviewed how to tell COM the names of our methods in an IDL file. Next we will explore how to define our argument lists and just what argument types COM can automatically pass between client and server.

Simple Argument Types

Although it's the intent of COM to allow a client to talk to any COM object, the reality is that some languages just don't work with other languages. The syntax of IDL is patterned most closely after C++ and it shows. For every argument type you can use in IDL you can also use as an argument type in your COM methods. But when it comes to Visual Basic and Visual J++ things start to fall apart.

In Table 2.2 is a list of all the basic argument types you can define in your IDL file and what to use for that argument type in three languages: C++, Visual Basic, and Visual J++. In other words, when you list an argument as type "boolean" in your IDL file, what argument type should you use in your Visual Basic program to receive this argument from a method call. (Trick question — VB doesn't support IDL's boolean argument type. You can still pass a boolean to VB, but as an IDL VARIANT. Use the "Boolean" type in VB and VARIANT_BOOL in C++.)

Note: Items marked as (1), etc., refer to notes at the end of the table.

Table 2.1   Simple Argument Types

What to send	IDL type	C++ type	VB type	VJ++ type
Unsigned 8 bits	char	char	Byte	char
Unsigned 8 bits	boolean	bool	(1)	(1)
Unsigned 8 bits	byte	BYTE	(1)	(1)
Unsigned 16 bits	unsigned short	unsigned short	(1)	(1)
Signed 16 bits	short	short	Integer	short
Unsigned 32 bits	unsigned long	unsigned long	(1)	(1)
Signed 32 bits	long	long	Long	int
Unsigned 64 bits	unsigned hyper	unsigned hyper	(1)	(1)
Signed 64 bits	hyper	hyper	(1)	(1)
32-bit floating point	float	float	Single	float
64-bit floating point	double	double	Double	double
Enumerator	Enum(2)	Enum(2)	Enum(2)	Enum(2)

What to send	IDL type	C++ type	VB type	VJ++ type
Currency (8 bytes)	CY	CY	CY	CY
Date (double)	DATE	DATE	DATE	DATE

NOTE 1: Although IDL might allow you to define a certain argument type, VB and VJ++ also depend on the COM DLL to know how to send that type. In this case, OLE32.DLL doesn't know what to do and you can't specify a proxy/stub DLL for the late binding that VB and VJ++ supports so this type isn't supported. What happens if COM realizes it needs a proxy/stub DLL but can't find it or load it? No, there's no error. COM just tries to communicate what it can and leaves the rest behind. In the case of an array, COM sends the first element only.

NOTE 2: Before you use an enumerator in your argument list, you must first define it at the top of the IDL file before any other declarations:

```
// enumerator (above any interface definitions)
typedef enum {Monday=2, Tuesday, Wednesday, Thursday, Friday} workday;
```

Then use it in your argument list like any other type:

```
method( [in] workday enumArg );
```

The enumerator's definition is automatically included any time you include this COM class definition.

Variable Attributes

Besides an argument type, in an IDL file you must also declare an argument's attribute(s). Fortunately there aren't that many to remember:

- [in] tells COM to only transmit this argument from the client to the server.
- [out] tells COM there's nothing to transmit to the server, but the server is going to have something to return back to the client. The client therefore *must* fill this argument with a valid pointer or NULL before it can call the method.
- [in,out] tells COM to not only send this argument to the server but to also return the data from the server back to the client. Again, the client must fill this argument with a valid pointer or NULL.

- `[out,retval]` these attributes tell the client that this argument should be returned as the result of the method call. In other words, it moves "value" out of the argument list to precede the method call as in this example:

```
value=method();
```

As you might expect, you can only use retval on one of your arguments and it must be the last one.

Wherever you use the "out" variable attribute, your argument type must also be a pointer:

```
…{out] long *pValue,…
```

Other variable attributes help to define variable sized arrays as seen next.

Arrays

When I say arrays, I mean the simple array of the form:

```
values[x];
```

where x is the number of elements in the array. Both IDL and the COM DLL support simple arrays. Therefore both C++ and VJ++ support simple arrays. But since the native type for an array in Visual Basic is a `SafeArray`, it does not. We will be revisiting `SafeArrays` later in this chapter.

You define an array as fixed or variable sized in IDL, and when variable sized, another argument in the argument list is called on to specify how big the missing size is. There's no way for COM to send an array if it doesn't know all of its sizes. Unlike a friendly call between methods in the same application, COM has to keep track of every byte to be sent. Otherwise it wouldn't know how many bytes to transmit.

Please refer to Table 2.3 for a list of variable attributes needed to declare an array in the IDL file and C++. For consistency sake, the table only shows arrays of longs being passed from a client to a server and where an array size is fixed, twenty-five (25) elements is used.

Table 2.2   Simple Array Argument Types

Type of Array	IDL declaration	C++ declarations
Fixed size	`long aArg[25]`	`long aArg[25]`
Variable sized	`[in,size_is(lSize)] long aArg[][25]`	`long lSize, long **aArg`
Variable sized/limited transmisson	`[in,first_is(lFirst), last_is(lLast), size_is(lSize)] long *aArg[][25]`	`long lFirst, long lLast, long lSize, long **aArg`
Variable sized/limited transmisson	`[in,first_is(lFirst), length_is(lLength), size_is(lSize)] long *aArg[][25]`	`long lFirst, long lLength, long lSize, long **aArg`

For fixed sized arrays, COM allocates exactly that much on the other side and transmits that much. For variable sized arrays, COM gets the number of bytes to allocate and transmit from another argument in the argument list. Note that only one dimension of an array can be variable.

For variable sized arrays, you can also optimize performance by telling COM to transmit only a certain number of elements, either by supplying the start and end element number or by supplying a start element and number of elements to follow. COM still allocates the full size of the array on the other side — it just doesn't transmit all of its elements.

For the syntax to use with VJ++, please refer to Example 35 (page 270).

Structures and COM Classes

Although IDL supports passing structures, the COM DLL does not. On the other hand, even IDL doesn't pass a regular C++ class — which is kind of inconsistent considering a structure in C++ is just another C++ class (a structure is a C++ class where the member variables and functions are by default public). But you can pass a COM class pointer.

To pass a structure using COM, you need to define that structure at the top of the IDL file as seen here.

```
// define structure above any interface definitions:
typedef struct
{
    long lElement;
```

```
   long *pPointer;
   float fElement;
} MYSTRUCT;
      :    :    :
   [in] MYSTRUCT myStruct
```

I

2

Note: If you notice, there's a pointer in this structure definition. Just as you need to make sure that all arguments in the argument list declared as [out] contain a valid pointer, so must you also make sure every pointer in a structure is valid.

As for COM classes, because all COM classes are derived from the IUn-known class, use that argument type or IDispatch for a COM class that supports late binding in the IDL file and type cast the pointer on the other side:

```
…,[in] IUnknown *iUnknown,…
   :    :    :
     IWzd *pWzd=(IWzd*)myClass;
```

Please see the next chapter for how to pass your COM pointer in a multitasking environment.

Encapsulated Unions

An encapsulated union allows you to put several different data types in the same union but pick at runtime what exact type represents so that COM can determine how many bytes to transmit. Variants operate along similar lines, however, an encapsulated union can be much more diverse. You can, for instance, unionize a structure and an integer. Encapsulated unions don't work with late binding so you're stuck with variants for VB and VJ++.

As with structures, you define an encapsulated union at the top of your IDL file like so:

```
// define above any interface definitions:
typedef [switch_type(long)] union //"switch_type()" makes switch a "long"
{
   [case(1)]
      float fFloat[2];
   [case(25)]
      double dDouble;
```

```
    [case(27)]
        MYSTRUCT myStruct;
    [default]
        long lLong;
} MYEUNION;
    :    :    :
    [in] long lType, [in,switch_is(lType)] MYEUNION myEUnion
```

In the client use:

```
MYEUNION myEUnion;
long lType=1;
myEUnion.fFloat[0]=123.0f;


pPtr->EUnions1(lType,myEUnion);
```

And in the server use:

```
STDMETHODIMP CWzd::EUnions1(long lType, MYEUNION myEUnion)
{
    if (lType==1)
        float f=myEUnion.fFloat[0];

    return S_OK;
}
```

Memory Pointers

When it comes to memory pointers, COM must not only transmit the pointer but also any memory the pointer was pointing to. It determines how many bytes to send based on how much memory is allocated to the pointer. Theoretically, a memory pointer can be pointing to any kind of memory. However, if there's the potential that the server is going to deallocate the memory that the client allocated, COM provides a set of memory API's called ::CoTaskMemAlloc() and ::CoTaskMemDealloc(). Not only do these functions keep memory from getting lost between objects, but they also create memory that's thread safe (two threads working with the same memory pointer can't write to the memory pointer at the same time).

There are three variable attributes you can apply to a memory pointer argument: ref, unique, and ptr:

- [ref] is the least overhead to COM. It just passes the memory pointer and any data it points to and the server promises not to change or deallocate the memory.

- [unique] is the next least overhead and the default for COM. This time the server *can* change the memory or even deallocate it, but the client and server promise that this is the only pointer to this particular chunk of memory. This way COM doesn't have to worry about reconciling what multiple pointers do to that chunk of memory while in the server before it sends it back to the client.

- [ptr] is the most overhead to COM, but the most transparent to you. You don't have to worry much about anything except making sure this argument is NULL if you're not using it.

Note: You can't use LPVOID to define a memory pointer because COM needs to know exactly how big of an argument you're sending. As with arrays, what used to be something you didn't have to worry about among the methods in your class, must now be carefully monitored. As an example, you can't simply pass a pointer to thousands of bytes of memory without expecting a performance hit, because COM has to send every one of those bytes to the server, which might be in San Francisco.

Visual Basic Argument Types

Before there was COM and IDL, there was Visual Basic. And some of Visual Basic's native argument types just don't fit into any COM category. These include how VB stores text strings, arrays, and untyped variables.

BSTRs

If you wanted to pass a character string using C++, you could use an array of characters. But to pass a character string to VB, you will need to use a binary string. Binary strings are composed of Unicode characters and are allocated with the Win32 API call ::SysAllocString(). In C++, you use a BSTR argument type to refer to a binary string argument. Or you can use a C++ helper class called _bstr_t.

SafeArrays

To pass just a simple array of numbers between C++ objects or even to VJ++, you would use the standard C++ array. But to pass an array to VB, you would use a SafeArray. A SafeArray, as its name implies, is more robust than a simple array, using a collection of Win32 API calls to allocate itself and prevent writing to more memory than you have. In C++, you would use a SAFEARRAY argument type to refer to a SafeArray.

Variants

Visual Basic is one of the few languages that still allows you to use a variable that isn't typed. To retrieve and pass such an argument to VB, you would use a Variant argument type. A Variant is simply the world's biggest union of just about every argument type there is. At any given moment, a Variant is just one of those types, such as an integer. But a Win32 API call will easily change it to another type if how the variable is used changes. As an example, if the variant that contained an integer was used in a text message, it would be converted to a BSTR. In C++, you would use a VARIANT argument type to refer to a Variant, or you can use a C++ helper class called _variant_t.

For more details on how to create and destroy these argument types using the API in C++ and classes in VJ++, please refer to Example 34 (page 265). Please refer to Table 2.3 for a list of how to specify these types in the IDL file and C++ and VJ++.

Table 2.3   VB Types

What to send/receive from VB	IDL type	VC++ type	VJ++ type
text strings	BSTR	BSTR or _bstr_t class	String class
arrays	SAFEARRAY(BYTE)	SAFEARRAY *	Safearray class
untyped arguments	VARIANT	VARIANT or _variant_t class	Variant class

Reverse Communication

Rather than tie up a client waiting for an event to occur at the server, it's much more efficient for the server to inform the client when something happens. There are several ways to do this yourself. One would be for your client to pass the server the address of a function to call when something happens. Another would be for a server to send the client a window message. But rather than implement either of these methods yourself, COM provides two standard ways for a server to communicate with its client that uses both of these solutions.

Connection Points and Sinks

In Win32 parlance, when one function passes an address to another function to call when some event occurs, that address is called a *callback address*. Since COM is brand new and spanky, new terms were required. And because one of the past analogies for COM was that objects were like hardware components, hardware terms were used. Thus the new terms *connection point* and *sink* were born. What those terms really mean is that one or more clients can give a server a callback address that the server will call when something happens. The address the server calls in a client is considered a sink and a server that can do this is considered to have a connection point.

Unfortunately, a client can't just pass a callback address to a server for all the same reasons why we weren't calling the server directly ourselves. Instead, the client itself must implement a mini-server COM server within itself and pass its pointer to the server to call. Please see Figure 2.2.

Figure 2.2   Connection Points and Sinks

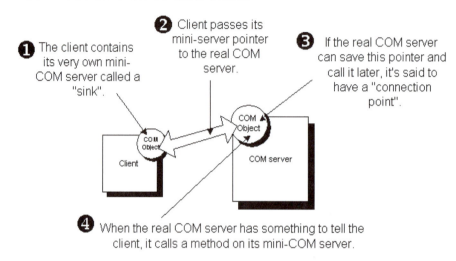

① The client contains its very own mini-COM server called a "sink".

② Client passes its mini-server pointer to the real COM server.

③ If the real COM server can save this pointer and call it later, it's said to have a "connection point".

④ When the real COM server has something to tell the client, it calls a method on its mini-COM server.

Another dimension to this solution is that just as clients and servers might be communicating through early and late binding interfaces, so must their connection points and sinks.

Early-Binding Interface

Since Visual Basic and Visual J++ only support a late-binding interface, you can't use early binding for its connection point and sink either. Only MFC and ATL support early-binding connection points and sinks, and surprisingly, it's almost as hard to implement in ATL as it is in MFC.

To implement a sink in an MFC client, you need to start by creating a COM interface project for that client with its own COM class and methods, IDL file and proxy/stub DLL. You then pick a client class that is derived from CCmdTarget — such as the main window (CMainFrame or CDialog) — and stick the same MFC macros in its .h file that you would use in a regular MFC COM server to implement a COM class:

```
DECLARE_INTERFACE_MAP()
    BEGIN_INTERFACE_PART(WzdSinkClass, IWzdSink)
        STDMETHOD_(HRESULT,Callback)(long);
    END_INTERFACE_PART(WzdSinkClass)
```

In this example, `IWzdSink` is the name of your client's COM class interface, `WzdSinkClass` is the implementation of that COM class interface and `Callback` is the method of that COM class.

You then implement `WzdSinkClass` in your client class as usual — it's just that now one of its methods is the callback function. Two more steps and we're done: create an instance of this COM class and pass it to the COM server supporting the connection point. Actually, you don't need to create an instance of your COM class because, as the main window of your application, it will already exist, so all you need to do is use the `QueryInterface()` function you just wrote to get a pointer to itself:

```
hr = m_xWzdSinkClass.QueryInterface (IID_IWzdSink, (void**) &iWzdSink);
```

To connect this sink to the connection point, you use MFC's `AfxConnectionAdvise()` static function, supplying the pointer we just got and a pointer to the COM object that supports the connection point. For a real example of an MFC sink, please refer to Example 14 (page 163).

A COM server that supports a connection point can be written in either MFC or ATL. If written in MFC, you use MFC's macros to add a connection map to any COM server class. In the `.h` file, you use these macros:

```
DECLARE_CONNECTION_MAP()
BEGIN_CONNECTION_PART (CWzdSrv, CallBackCP)
    CONNECTION_IID(IID_IWzdSink)
END_CONNECTION_PART(CallBackCP)
```

And in the `.cpp` file you use these macros:

```
BEGIN_CONNECTION_MAP(CWzdSrv, CCmdTarget)
    CONNECTION_PART(CWzdSrv, IID_IWzdSink, CallBackCP)
END_CONNECTION_MAP()
```

where `CWzdSrv` is the regular COM server class, `IID_IWzdSink` is the Interface ID of the client's sink mini-server, and `CallBackCP` is the name of a new member variable of `CWzdSrv` declared with MFC's `ConnectionPoint` class and called `m_xCallBackCP`. This class not only keeps track of the address of your client's sink, but also any client that connects to this server. However, you have to implement the method that actually calls the client back. This you do in a new member function of your COM class like so:

```
void CWzdSrv::CallBackClients(long data)
{
   const CPtrArray *pConnections=m_xCallBackCP.GetConnections();
   int nConnections=pConnections->GetSize();
   for (int i=0;i<nConnections;i++)
   {
      IWzdSink *pWzdSink=(IWzdSink*)(pConnections->GetAt(i));
      pWzdSink->Callback(data);
   }
}
```

Notice how you use the new member variable m_xCallBackCP to get the number of clients and the addresses of their sinks. For a real example of an MFC connection point, please refer to Example 13 (page 160).

If you write your connection point in an ATL COM server, the ATL Object Wizard will add the connection map for you, however you still need to fill it in manually and add the routine that will be calling the clients back manually as well. To ask the wizard to add the connection map, you select "Supports Connection Points" when creating the object. This causes the wizard to generate the map in your COM class's .h file. You then need to fill this map in yourself like so:

```
BEGIN_CONNECTION_POINT_MAP(CWzd)
      CONNECTION_POINT_ENTRY(IID_IWzdSink)
END_CONNECTION_POINT_MAP()
```

where IID_IWzdSink is again the Interface ID of the client mini-server that's supporting the sink. The callback routine you need to add looks very similar to the MFC routine, except that it exists in its very own class derived from IConnectionPointImpl:

```
template <class T>
class CProxyIWzdSinkEvents : public IConnectionPointImpl<T,
   &IID_IWzdSink, CComDynamicUnkArray>
{
public:
   HRESULT Fire_WzdSink( long lArg )
   {
```

```
        T* pT = (T*)this;
        pT->Lock();
        HRESULT ret;
        IUnknown** pp = m_vec.begin();
        while (pp < m_vec.end())
        {
            if (*pp != NULL)
            {
                IWzdSink *pWzdSinks = reinterpret_cast<IWzdSink*>(*pp);
                ret = pWzdSinks->Callback( lArg );
            }
            pp++;
        }
        pT->Unlock();
        return ret;
    }
};
```

This class becomes one of the many root classes of your ATL COM class so that you can again call it from anywhere in your COM server. For a real example, please see Example 25 (page 220).

Late-Binding Interface

With early binding, we created a whole new interface project for our client that included an IDL file and a proxy/stub DLL to define the mini-COM server that will support the sink. Where then do we stick the IDL file for our late-binding clients' Visual Basic and Visual J++? COM's solution was to define it in the server with the connection point. And if you think about it, that makes a lot of sense. It's not like you can use a connection point with just any sink anyway. They must each support the same unique methods and argument lists. So why not join their definitions together in the same IDL file?

To add a connection point to an ATL server, you start by specifying "Supports Connection Points" in the ATL Object Wizard. Only this time, rather than manually adding your own methods to the connection map, you add the sink methods just like you would any ATL method through the ClassView. The Studio not only automatically adds the connection point automatically to your connection map, but it also sticks a definition for the client's mini-server into the IDL file that's stored in the type library:

```
library SERVERLib
{
    importlib("stdole32.tlb");
    importlib("stdole2.tlb");

    [
        uuid(37D9335E-A706-11D3-A398-00C04F570E2C),
        helpstring("_IWzdEvents Interface")
    ]
    dispinterface _IWzdEvents
    {
        properties:
        methods:
    };

    [
        uuid(37D9335D-A706-11D3-A398-00C04F570E2C),
        helpstring("Wzd Class")
    ]
    coclass Wzd
    {
        [default] interface IWzd;
        [default, source] dispinterface _IWzdEvents;
    };
}
```

In this example, `dispinterface` indicates this is a late-binding interface, _IWzdEvents is the name of the client's mini-server, and the interface attribute `source` tells the MIDL compiler that the client will be implementing this interface rather than the COM server. Please refer to Example 25 (page 220).

So the definition of the client's sink now sits in the COM server's type library waiting for some client to read it and implement it. When you pull this type library into a Visual Basic client, the VB Studio reads this definition and automatically allows you to add the sink's methods and argument list to any form using the COM server. Please see Example 28 (page 236).

For a Visual J++ client, the definition for the sink is pulled out of the type library when you ask the VJ++ Studio to wrap a COM server with a Java class. You can then use "`implements`" to implement the sink in your VJ++ classes. Please see Example 30 (page 240).

ActiveX Events

The ActiveX standard takes an entirely different approach to server to client communication. Instead of directly calling a client method, ActiveX uses the approach of sending the client a window message. In an MFC client, this message is processed in an event map you create with the ClassWizard. In VB, you're totally removed from the process and simply use the VB Studio to add an ActiveX's events to your forms.

Summary

Although COM does its best to allow any COM object to communicate transparently with any other COM object, the native data types of each computer language make this an impossible goal. You could always program for the least common denominator (Visual Basic), but to write the most optimized code, you need to carefully evaluate what COM object will be talking with whom and why. And the tables in this chapter should hopefully help keep this Tower of Babel together.

Now that we've covered what goes into a COM object and how they communicate with each other, there remains the subject of everything else. Can or should COM classes derive their functionality from one another? How do you implement licensing for your work when your work is a COM object that's part of another application? How will your COM object behave in a multitasking environment?

Chapter 3

Other COM Issues

So far, we've talked about how to create a class object from any application and how to talk to it. But once you implement such a technology, you find a lot of side issues come up. Such as how do you keep just anybody from creating that object, especially if it accesses sensitive data? Or how will that object behave in a multitasking application? Or should you allow an object to save its data between invocations? Or can you design one object to derive some of its functionality from some other COM object "base" class?

In this chapter, we will find that when it comes to security, creating an object on another machine is not much different than if you were to go to the machine yourself and run a program there. We will find that COM also has a built-in standard that requires an application to have a license to run your object. We will discover that COM can be used to automatically make your object thread safe in exchange for some CPU time. But we start by reviewing how you would go about deriving functionality from some other COM object in your own.

Encapsulation and Aggregation

Wouldn't it be fantastic if you could derive new functionality from existing COM classes on your system? You could build powerful, potentially bug-free applications from all that existing work. Unfortunately, what is traditionally considered inheritance — the ability for one class's methods to bleed through and be automatically available in the derived class — is not currently available with COM. Why not? Traditional inheritance is achieved at compile-time when a virtual table (VTBL) is created to all of the methods in the class as seen in Figure 3.1.

Figure 3.1   Implementing Traditional Inheritance

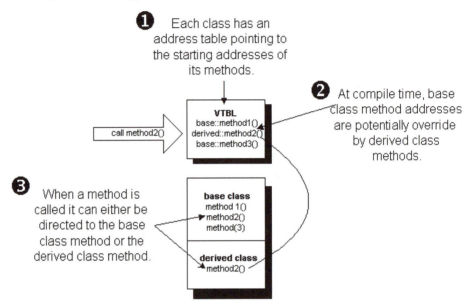

COM classes can't interact at compile-time because they might be written in six different languages and those languages may not even have a VTBL. Also consider that your base class's VTBL might be in Cleveland while your derived class's VTBL could be in Cincinnati — not exactly contiguous.

Therefore, any support of traditional inheritance in COM would have to be a runtime solution, and that runtime solution would involve overhead and complexity and a lot of headache. An example of a runtime solution might go like this: you create a COM object IWzd2 with a base class IWzd. When you go to call a method on IWzd2, COM must first check to see if the method you're calling is actually in IWzd2 or if it's in fact a method in IWzd and then make the call there.

Microsoft has been promising to eventually add runtime inheritance to COM, but it probably won't be soon considering how long they've promised it. Instead they recommend and support one of two other methods: encapsulation or aggregation.

Encapsulation

With encapsulation, you're simply creating an instance of a COM "base" class from within your "derived" COM class and passing any method calls directly to the base class. You create the base class in your derived class's constructor using COM and release it in your destructor. The really hard part is wrapping all of the base class's methods with your derived class. With MFC, this would look something like this:

```
STDMETHODIMP CWzdSrv::XWzdClass::Method2(long lArg, unsigned long ulArg)
{
    METHOD_PROLOGUE(CWzdSrv, WzdClass);
    return m_iBaseClass->Method2(lArg,ulArg);
}
```

Not only is this tedious, but you're prone to make mistakes, or at least more mistakes than with traditional inheritance. Please see Figure 3.2.

Figure 3.2   Encapsulation

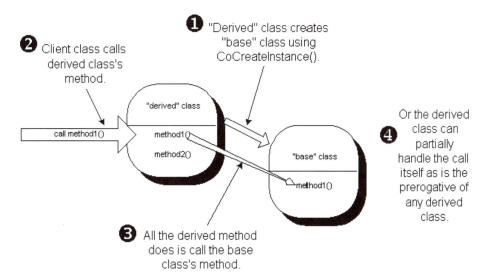

❶ "Derived" class creates "base" class using CoCreateInstance().

❷ Client class calls derived class's method.

❸ All the derived method does is call the base class's method.

❹ Or the derived class can partially handle the call itself as is the prerogative of any derived class.

Aggregation

Another technique called aggregation starts out just like encapsulation, except instead of you manually wrapping every method, your client must call QueryInterface() from your derived class to get an interface to your base class's methods. Effectively for your user, you've just replaced their call to CreateInstance() with a call to QueryInterface() for the base class. So with aggregation you've delegated all the work to your client and eliminated the possibility of introducing bugs into your own code. But it's not really inheritance is it? What you do get — other than the automatic creation of two COM objects at once — is the assurance that both objects are in the same state, and your derived class can control your base class. Aggregation is used in the next chapter for transactions in COM+ for this reason. Please see Figure 3.3.

Figure 3.3 Aggregation

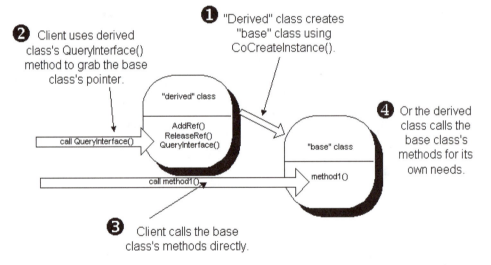

Another positive side to aggregation is that ATL and MFC both provide support for it. In MFC, you create the base class as usual, except you call EnableAggregation() in its constructor. In the derived (aggregating) class, you use the ClassWizard to override CCmdTarget's OnCreateAggregates() member function and create the base class there using COM as usual. If it fails, you return false and the derived class won't be created. You also release the base class in your OnFinalRelease() member function and stick a macro in your interface map so a `QueryInterface()` will work. For all the details, please see Example 16 (page 173).

With ATL, you create the base class as usual using the ATL Object Wizard, making sure to select the option that will allow the object to be aggregated. In the derived (aggregating) class you create the base class in its `FinalConstruct()` and release the base class in its `FinalRelease()`. Once more, you add a special ATL macro to the interface map so that a client using `QueryInterface()` will be able to get a pointer to this base class. For all the details of using ATL aggregation, please refer to Example 26 (page 226).

As for how your client will access this base class, smart pointers make it fairly easy. If you've already created a COM object using `CreateInstance()` and you want to create a smart pointer to access the base class, you can use:

```
IBaseClassPtr pBaseClass(pDerivedClassPtr);
```

where `IBaseClassPtr` is the base class smart pointer. Internally, the smart pointer is still doing a `QueryInterface()`, but now you don't have to write and test it. Be ready to catch exceptions though if the smart pointer can't find the base class's interface. Your client's application might then look something like this:

```
pBaseClass->Method1(1234,&lArg);
pDerivedClass->Method2(lArg2, ulArg);
```

Even though one object is aggregated from the other, you must still release both interfaces. However, with smart pointers, you don't need to worry.

Security

Another topic of concern to COM writers is security. Your COM classes can have access to some very secure data sources, so you wouldn't want just anybody to be able to create and use them. For objects running on your own system, COM relies on whatever security you decide to enable using Windows — the assumption being that if you have access to a system and can log into it, then you have enough privilege to use its software.

What about users on other systems in your network? How do you prevent them from creating and using the COM objects on your system using DCOM? In this case, COM provides two types of security: Activation Security and Application Security.

Activation Security

The first type of security, called Activation Security, simply prevents a COM object from being created by a user on any other system. To do this, you change a setting in your system's registry, either using the Registry Editor (regedit.exe) or OLE/COM Object Viewer (oleview.exe). Please refer to Figure 3.4 for how to turn off DCOM access to your machine using OLE/COM Object Viewer.

Figure 3.4   Deactivating Remote Access to Your COM Objects

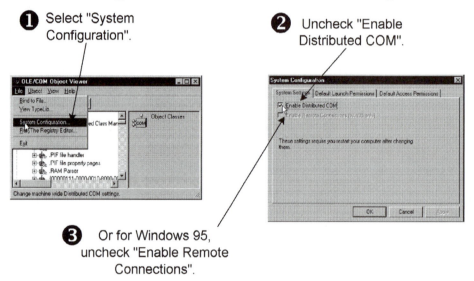

❶ Select "System Configuration".

❷ Uncheck "Enable Distributed COM".

❸ Or for Windows 95, uncheck "Enable Remote Connections".

Note: Even if you deactivate DCOM to prevent outside access to your system, your system can still use DCOM to access another machine.

Object Security

The second type of security allows you to be more selective about what objects can or can't be created and by whom. Actually, the first hurdle to a remote user accessing your COM objects on a Windows NT/2000 system is that they need to have an account on your machine. This account must be identical to the one with which they logged into their own system. In other words, if they logged into their machine using the User ID and Password of

"Smith" and "12345", then they must have the identical User ID and Password on your system. That's because when COM goes to create an object on your system, it firsts automatically logs into your system as that individual.

The second hurdle a remote user must pass is a security check you add to each COM object using the OLE/COM Object Viewer. With this utility, you can tell a COM object just which users are authorized to access it. Security is configured in two parts: **Launch Permissions** determines which users are able to create the COM server and its objects, and **Access Permissions** determine which users can access the object once it's been created. This latter security is important if the object's pointer is passed to a user on a third machine. Launch Permissions should always be more restrictive then Access Permissions — at least according to the documentation. My philosophy would be that you shouldn't be able to access an object if you can't create it, so I always make sure the settings are the same.

Security permissions for a COM object are stored with its other settings in the system registry. You can access these settings using the OLE/COM Object Viewer as seen in Figure 3.5.

Figure 3.5   Setting Security Permissions for a COM Class

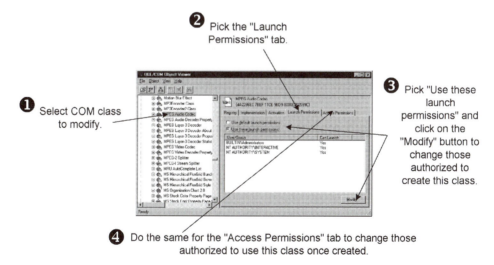

❷ Pick the "Launch Permissions" tab.

❶ Select COM class to modify.

❸ Pick "Use these launch permissions" and click on the "Modify" button to change those authorized to create this class.

❹ Do the same for the "Access Permissions" tab to change those authorized to use this class once created.

CoInitializeSecurity()

An even finer granularity of security can be achieved for your object pro-
grammatically using CoInitializeSecurity(). COM uses this same API
call itself with your system registry settings as its input. But if you call this
function first, before the first out-of-process method has been called, you
can override these settings. Unfortunately, this API call is rather complicated
and really reserved for a truly advanced user. And since your settings in the
system registry will take care of 99% of your security needs, why bother?

Note: Even finer granularity can be achieved if your object uses
COM+. As we will discover in the next chapter, you can even assign
security to the individual methods of a COM object.

Licensing

A topic that's closely related to security is licensing. Again, you're limiting
access to a COM object but for a different reason ("I don't care about your
nuclear secrets, I just want my money."). A lot of documentation has been
devoted to an interface called IClassFactory2 that standardizes how to prevent an object from running on a machine without a license. However, as
with any standard, talk is cheap; so how do I get my project done on time?
After all, you can add your own proprietary licensing protection scheme to
your COM object which would probably be tougher to crack then if it were
to follow some well-published approach.

On the other hand, both MFC and ATL do provide some support for
licensing that you might want to take advantage of — although, as usual,
MFC is harder to use. That's because you need to override a member function of an MFC class that's hard coded into a macro. However, once you
follow what to do in Example 44 (page 324), it's a simple matter to add
logic that checks for a license. In this example, we use MFC's own
AfxVerifyLicFile() function to look for a license file on a disk and compare it to what it should be:

```
virtual BOOL VerifyUserLicense()
{
    return AfxVerifyLicFile(AfxGetInstanceHandle),
```

```
            "licence.lic",
        L"1234567890",
            -1);
}
```

Of course, if your user is crafty enough to copy your COM object onto a floppy, they're probably also going to copy the license file too — so you may want to hide it somewhere or require it be refreshed or check against something else instead. But the effect you're going for is that if this is an unlicensed machine, you want to return a FALSE from VerifyUserLicense().

ATL is somewhat easier to use. As shown in Example 45 (page 329), you don't have to tear apart some ATL macro to have it use your licensing logic. Instead, you declare it with DECLARE_CLASSFACTORY2(CWzdLicense).

You still have to manually add the class that supports this logic, which can look very similar to the MFC code:

```
class CWzdLicense
{
protected:
    static BOOL VerifyLicenseKey(BSTR bstr)
    {
        // compare bstr with embedded license info
        return TRUE; // valid license
    }
}
```

Multitasking

So how does your COM object behave when created from a thread other then your main process? The simple answer is: just fine. Other then trying to remember to initialize the COM DLL from every thread in which you want to use a COM object, your object will behave no differently than if it were a non-COM object. The trouble starts however in that COM has decided to get into the thread safety business. In other words, at the same time that COM creates your object and lets you talk to it, in certain circumstances, it will also synchronize your communications with it so that no other thread can talk to it at the same time. Why is that required in a multitasking environment?

Thread Safety Primer

Although adding a thread or two to your application allows you to process data simultaneously with your main process, it also adds a new potential problem to your design — namely, how do you keep two threads from modifying the same global resource, such as a data table, at the same time? As an example, if both your main process and a thread have access to the same global table g_Table and just as your main process was writing data to it, the thread was reading data from it, the data that the thread would read could be half new data and half old data — in other words, corrupted data.

The standard solution to this problem involves setting up a global "occupied" flag which the main process and each thread checks and sets before it accesses the resource. If the resource is busy with another thread, the new thread waits its turn, possibly timing-out rather then accessing that resource. With Windows, you can use the EnterCriticalSection() and LeaveCriticalSection() API functions to setup and check this "occupied" flag.

COM Thread Safety

COM offers two solutions beyond critical sections for thread safety. The first solution is based more on what you're expected to do than what COM will do for you. Essentially, you must promise never to access a shared resource from more than one thread ever. The steps to this solution are as follows:

1. Identify your shared global resources, such as a table called g_Table;

2. Identify the COM objects that read and write to g_Table;

3. Determine which thread or main process will most often use these COM object(s);

4. Only create these COM object(s) and call their methods from that thread/main process;

5. Never pass a pointer to one of these created objects to another thread/main process, so that it can call those methods.

And that's it. COM does nothing for you. Please see Figure 3.6.

Figure 3.6   Manual Thread Safety

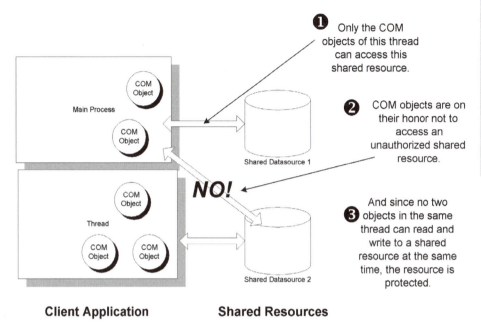

Client Application   **Shared Resources**

The other COM solution is much more proactive.

Automatic COM Thread Safety

You can also configure your COM objects so that they *cannot* access a shared global resource from more than one thread. From the previous example, you can configure your COM objects so that even if they were created in any thread or the main process, the created objects would be incapable of running at the same time because COM would automatically create them all in the same communal thread. Please see Figure 3.7.

Figure 3.7   **Automatic Thread Safety**

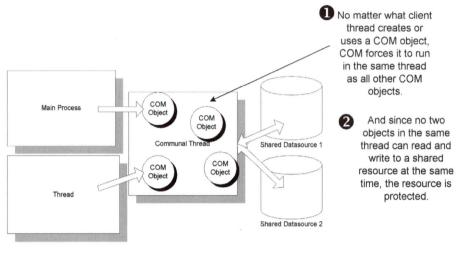

❶ No matter what client thread creates or uses a COM object, COM forces it to run in the same thread as all other COM objects.

❷ And since no two objects in the same thread can read and write to a shared resource at the same time, the resource is protected.

Client Application   **Shared Resources**

So how does COM force an object to run in a thread different from its caller? Actually, it's pretty clever. Instead of calling a method directly, your call is converted into a message and that message is sent to a message queue in the communal thread. The communal thread then processes these messages one-by-one at its own leisure. If two or more threads are using COM objects that access a share resource, the first thread to call a COM object is processed while the other thread's calls sit waiting for a response from their messages. In other words, COM is using message loops to synchronize access to the shared resource.

This protection kicks in whenever you first create your COM object. Without any protection, the pointer returned to you from COM simply points to a VTBL of methods available from that COM object. With protection, the returned pointer points to a proxy routine that does the message loop waiting.

If any of this sounds familiar, that's because a similar technology is also used when your application makes out-of-process method calls to an EXE server. What's different is that simple message queues are used for marshalling instead of the Remote Procedure Call (RPC) API. You still need a custom proxy/stub DLL if you are passing custom data. Although passing data within a process is less CPU-intensive then out-of-process, your data is still man-handled onto the thread's stack — let's just hope they optimized as best they could.

Please see Figure 3.8 for a review of automatic COM thread safety.

Figure 3.8   Thread Safety Through Message Loop Synchronization

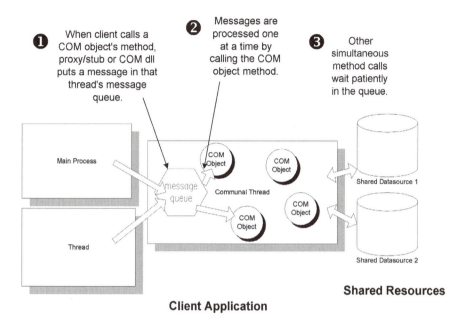

In essence, this solution is the opposite of multitasking — instead of creating new threads to handle multiple tasks all at the same time, COM is creating one new thread to handle all of the same type of task one-at-a-time. Hopefully, the drawback is also apparent. Waiting on an entire method to finish, instead of a smaller critical section, can add significantly to your throughput — not to mention the overhead added by converting and unconverting your method call into a message.

Activating Thread Safety

So how do you tell COM, "Yes, I too would like to share in the bounty of COM and have my objects protected. Please."? And how do you turn it off? Unfortunately, you just can't flip a switch. In fact, whether or not your object is protected depends on several factors, including:

- what "concurrency model" you specify when initializing the COM DLL. In other words, when calling `::CoInitializeEx()`, do you use `COINIT_APARTMENTTHREADED` or `COINIT_MULTITHREADED`?

- what threading model you specify for your objects: Single, Apartment, Free or Both;
- whether or not your COM objects reside in a DLL or an EXE;
- and whether or not your COM object is being created by another COM object.

As you can see, the permutations can get ugly fast, which is probably why Microsoft came up with an apartment analogy (more on that later). But to really get a feeling for what's going on, let's go through the permutations setting by setting. Tables 3.1 and 3.2 show what happens when you create COM objects from in-process (DLL) servers. Tables 3.3 and 3.4 show what happens with out-of-process servers (EXE) whether local or remote. Tables 3.1 and 3.3 show what happens when you use the `COINIT_APARTMENTTHREADED` concurrency mode when initializing COM and Tables 3.2 and 3.4 show the same for `COINIT_MULTITHREADED`. As an example of reading these tables, the first entry in Table 3.4 would read, "The main process creating an object with the Single, Apartment, or Both threading model would run this object in the main process's thread and any calls to its methods would be directly to that object's VTBL — not through a message queue."

Table 3.1   In-Process (DLL) Server Using `COINIT_APARTMENTTHREADED`

This Entity...	Creating an object with this threading model...	Creates the new object in this thread...	And method calls to the object are....
The main process	Single/Apartment/Both	Main process	Direct
	Free	New	Synchronized
A thread	Single	Main Process	Synchronized
	Apartment/Both	Same	Direct
	Free	New	Synchronized
Single object	Single/Apartment/Both	Same	Direct
	Free	New	Synchronized
Apartment or Both object running in main process	Single/Apartment/Both	Same	Direct
	Free	New	Synchronized

Apartment or Both object running in a thread	Single	Main Process	Synchronized
	Apartment/Both	Same	Direct
	Free	New	Synchronized
Free object	Single	Main Process	Synchronized
	Apartment	New	Synchronized
	Free/Both	Same	Direct

Table 3.2   In-Process (DLL) Server Using `COINIT_MULTITHREADED`

This Entity...	Creating an object with this threading model...	Creates the new object in this thread...	And method calls to the object are....
The main process or thread	Single/Apartment	Communal	Synchronized
	Free/Both	Same	Direct
Single object	Single/Apartment/Both	Same	Direct
	Free	New	Synchronized
Apartment object running in main process	Single/Apartment/Both	Same	Direct
	Free	New	Synchronized
Apartment object running in a thread	Single	Main Process	Synchronized
	Apartment/Both	Same	Direct
	Free	New	Synchronized
Free/Both object	Single/Apartment	Communal	Synchronized
	Free/Both	Same	Direct

Table 3.3   Out-of-Process (EXE) Server Using
`COINIT_APARTMENTTHREADED`

This Entity...	Creating an object with this threading model...	Creates the new object in this thread...	And method calls to the object are....
Any External Client	Any	Communal	Synchronized
Any COM object (Single,etc.)	Any	Same	Direct

Table 3.4   Out-of-Process (EXE) Server Using
`COINIT_MULTITHREADED`

This Entity...	Creating an object with this threading model...	Creates the new object in this thread...	And method calls to the object are....
An external Client	Any	New	Synchronized
Any COM object (Single,etc.)	Any	New Communal	Synchronized

The threading keywords in these tables mean the following:

- **Main process** — the object runs in the main process. This is the slowest of the alternatives because each object not only must fight for CPU time with other similar objects, but it must also share time with whatever goes on in the main application. However, it also means any global resource these objects access can also be accessed by the main process.

- **Same** — the object runs in the whatever thread created it, whether that be the main process or an application thread.

- **New** — a new thread is created for the object to run in. This is used with Free threaded objects to allow them to execute concurrently with their creator threads, hopefully allowing for greater throughput. This is the one case where COM creates more threads for faster execution then less threads for thread safety.

- **Communal** — a thread created for all objects of this type to run in. All objects in this commune are therefore protected from interfering with its companion objects when accessing a global resource.

Other Important Notes:

1. You must call `::CoInitialize()` or `::CoInitializeEx()` from *each* thread your application creates before you can use COM from that thread!

2. When using the older `::CoInitialize()` call, the concurrency model defaults to `COINIT_APARTMENTTHREADED`.

3. With out-of-process servers, it doesn't matter what concurrency mode your client application uses — it only matters what the server uses.

4. The threading models (e.g., Single, Apartment, etc.) are irrelevant for out-of-process servers. COM only makes use of this setting (which is stored in the system registry) with in-process servers.

5. There's a bug with the Single threading model: if you create a Single threaded object from a thread before you create one from the main process, the Single object will be created in a new thread instead of the main process. Then, when attempting to create one from the main process, it will fail. But because the Single model is no longer supported, this bug will probably never be fixed.

Synchronization Loopholes

If while perusing Tables 3.1 to 3.4 you find yourself wishing you could call a method directly instead of going through the message queue, or conversely, force your call to go through the message queue so you can safely break into an apartment, COM provides three API calls to do just that.

- `CoCreateFreeThreadedMarshaler()` allows you to get a direct pointer into a COM object no matter what the original configuration. In Examples 47 and 48 (beginning on page 337), we actually aggregate our COM object to the COM object created by this API call. In COM terminology, you've replaced your regular marshaller (as supplied by the COM DLL or your own proxy/stub DLL) with a "Free Threaded" marshaller. And this free-thread marshaller, rather then marshalling anything, simply directs your calls right to their methods.

- `CoMarshalInterThreadInterfaceInStream()` and `CoGetInterfaceAndReleaseStream()` convert a direct pointer into a synchronized pointer while taking a short excursion in a stream created just for the occasion — although not always. In fact, it's up to COM whether or not to synchronize the pointer based on all of the factors listed above (threading models, etc.) If, for instance, you are passing a COM pointer between two objects in the same thread, the pointer you extract on the other side won't be synchronized. Please see Example 32 (page 260).

Designing Your In-Process COM Server

Tables 3.1 to 3.4 are all well and good for trying to make sense of an existing configuration. However, how should you design a new application?

If you want to keep COM from doing anything, use a concurrency mode of `COINIT_MULTITHREADED`, and use the "Free" model for all of your COM objects. You must then provide your own thread safety, possibly by putting critical sections around all data access to a shared resource.

A way to prevent COM from doing anything to make your object threads safe (but at the same time not having to use critical sections) is to use COM's strategy of running all COM objects that access a particular shared resource from the same thread. COM, again, does nothing but you have to be very careful about what object accesses what. And, of course, you lose out on the benefits of multitasking for any logic in those objects. So if you want to be on your honor to prevent objects in two threads from accessing the same global resource, you set the concurrency mode to `COINIT_APARTMENTTHREADED` and each COM object to use "Apartment" threading.

To have COM take over your object's thread safety, use a concurrency mode of `COINIT_MULTITHREADED` and the same "Apartment" threading model for each of your objects. All of your objects, no matter what thread they were created in, will be forced to run in the same thread and your access to it will be synchronized with any other thread's access to an object within. This is obviously the slowest solution but the safest.

The danger in the fastest solution of using critical sections is that you might miss a spot to put a section leading to a flaky application. However, if you do plan to use this approach and encounter flakiness, you can still use COM to isolate the problem. Change all of your COM objects to the "Apartment" threading model — essentially asking COM to ensure your objects are safe. (If the flakiness persists, your problem is just shoddy programming.) Then one-by-one convert them back to "Both" until the problem reappears. As you can see from Table 3.2, the "Both" threading model will not run in the communal thread. (You knew there had to be a use for "Both"!)

Designing Your Out-of-Process COM Server

The settings for an out-of-process (EXE) server are dramatically simpler but also less flexible. Because it's out-of-process, all objects are by default running in another thread with their pointers synchronized. So there are really only two choices:

1. If you don't want to use critical sections, set the concurrency mode of the server (*not* YOUR CLIENT APPLICATION) to COINIT_APARTMENTTHREADED. All objects will then run in the same thread.
2. When desiring faster throughput, use critical sections and COINIT_MULTITHREADED. All objects will then run in their very own thread created by COM.

If there's a flaky problem with solution (2), change your concurrency mode to COINIT_APARTMENTTHREADED.

Apartments

Probably the most complicated part of COM threading is the terminology that Microsoft uses to describe it. To this end, they came up with the concept of COM objects living in an apartment. If you thought the whole topic of threading required some abstract thinking, an apartment is worse.

As mentioned earlier, one of the ways COM uses to automatically make your objects thread safe is to force them to all run in the same thread and then synchronize any access to that thread using a message loop. By synchronizing access, any thread calling a method of an object in the protected thread has to wait until any other thread has finished their call, even if it's to another method on another object. This protected thread is called a Single Threaded Apartment (STA) — providing thread safety for any COM object living inside.

If you decide to provide your own thread safety in your COM object by using critical sections, any thread at any time can access the methods of your object. Collectively, all of the COM objects that allow this asynchronous access are considered to be living in a Multithreaded Apartment (MTA). Because that means *all*, there's only one MTA in an application.

So you might be going along thinking, "ah, an apartment is just a thread with synchronized access to the functionality inside" until you come across a multithreaded apartment which is really just a rag-tag collection of objects that can fight for themselves — an anti-apartment really. Please see Figure 3.9 for more, but if this still doesn't help, you really don't need to know what an apartment is to be proficient with multitasking COM objects. Because no one really seems to know what these words mean anyway, you can make it up as you go along in conversation, "Does this object share an apartment with your free-threaded model? Because maybe you should consider using a "Both" object instead — no, I have that backwards."

Figure 3.9   Applications Can Be Composed of STAs and MTAs

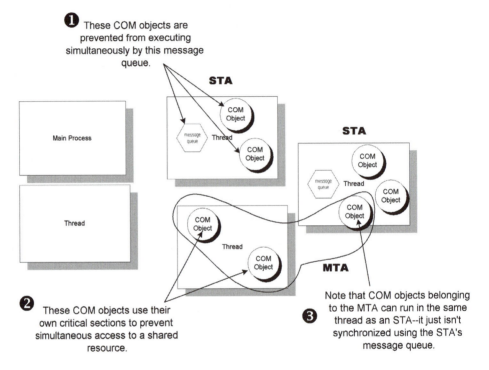

❶ These COM objects are prevented from executing simultaneously by this message queue.

❷ These COM objects use their own critical sections to prevent simultaneous access to a shared resource.

❸ Note that COM objects belonging to the MTA can run in the same thread as an STA--it just isn't synchronized using the STA's message queue.

Summary

In this chapter, we found that COM provides two weak forms of class inheritance where you do most of the work. **Encapsulation** is the concept of wrapping a "base" class's methods with your own "derived" class. **Aggregation** is the concept of your derived class starting up a base class and then exposing its methods, although a client still has to use QueryInterface() to get to them.

We also looked at the facilities that MFC and ATL provide for supporting licensing of your COM objects, and at Windows NT/2000 support for security. Finally, we looked at using COM to make your objects thread safe and almost blacked out trying to understand what an apartment is.

In the next chapter, we will be entering the wonderful world of COM+, which rather than what the name infers, is really an enhancement of just one aspect of COM — an aspect that we really haven't even discussed yet: using DLLs out-of-process.

Chapter 4

COM+

A while ago, when I was first reading the literature on COM, I ran across an obscure article on DLL surrogates. Until then, I thought that COM DLL servers could only be used in-process, and that if you wanted to develop a server on another machine, you had to use a COM EXE server. A DLL surrogate, I read, could now allow you to run a DLL remotely by loading the DLL for you and relaying the object pointers to you. At the time, I thought it was an oddity developed just for completeness sake. Microsoft themselves recommended it to be used on flaky DLLs so that they couldn't crash the main application. Little did I know it would be the basis for a whole new way to program.

Somewhere along the line someone realized that rather than have this dumb DLL surrogate just sit there relaying object pointers, it could also actually add functionality to your COM DLL without you changing a single line of code.

In this chapter, we'll review how the design of client/server applications evolved over the years and how COM DLL surrogates were used to help with that evolution. We'll watch surrogates evolve from the super dumb DLLHOST application, to the transaction savvy Microsoft Transaction Server, to the full service COM+.

We'll watch as the systems programming (e.g., thread safety, load balancing, scaling) that an application programmer used to have to worry about be reassigned back to the system where an application programmer can activate it by setting the right option.

The Evolution of Client/Server

As we've seen in previous chapters, COM takes a lot of the drudgery and headache out of writing client/server applications. No longer do you need to come up with your own protocols and talk to Window Sockets and essentially become a systems programmer when all you wanted to do is allow clients on one or more machines to share the resources and functionality of a common server.

But just as COM was being initially developed, the whole concept of client/server was changing. Some would say it was a revolution, but the old-timers insist it's just come full circle.

Back in the seventies and earlier, a classic client/server configuration would have had just one super-fast (16 bit, 6 Mhz) server with client dumb terminals accessing its programs and data. If this is before your time, all a dumb terminal knows how to do is print characters on its screen and transmit characters back to the server. But because everyone was sharing the same supercomputer (10MB disks), throughput could be deadly slow at peak hours.

Then along came PCs and suddenly everyone could have a little mini-computer on their desk that could run just their own applications with no peak hours to dread. And without being loaded down running every word processor program in the company, the server was much faster at what it should really be doing: sharing data with its clients.

But as client and server computers got more powerful, even more configurations became possible. As an example, one of the problems that came up when all of the word processor applications migrated out to the client was that upgrading to a new word processor became a nightmare for the systems administrator. In the bad old days, he/she would just upgrade the word processor on the server. But rather than stick this application back on the server and tie it up again, a three-tiered design evolved — with the clients still on one machine each talking to a server that runs the application, which in turn, talks to the server with all the shared data on it. Please see Figure 4.1.

Figure 4.1   Three-Tiered Architecture

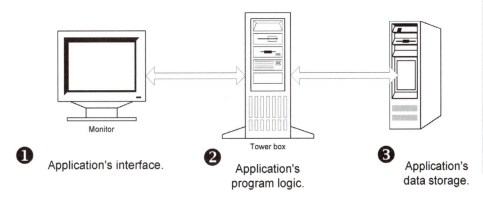

Monitor

Tower box

❶ Application's interface.

❷ Application's program logic.

❸ Application's data storage.

New terminology was required of course. Because the client was getting stripped of everything but an application's interface, it now became "thin client". And the stuff that was getting stripped out of the client was called its "business logic".

As server computers get even faster, the communication between application server and database server becomes a bottleneck. This is a particular problem when an application does a lot of cumulative database access, where it can't just grab all the information it needs at once, but has to sequentially grab data based on a previous action. Network communications can start to burn up the lines. Sure, stored procedures help by keeping some sequential accesses on the database server, but you don't want to write all of your application in stored procedures. So when the server is fast enough to handle the load, the application and database servers are being consolidated onto one machine in what's called a *two-tiered architecture* — or as someone from the seventies would call it, a classic client/server architecture. Please see Figure 4.2.

Figure 4.2　Two-Tiered Architecture

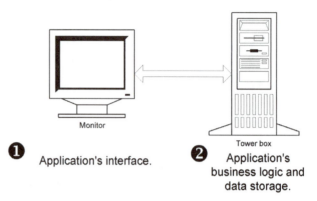

Monitor

❶ Application's interface.

Tower box

❷ Application's
business logic and
data storage.

In about ten years or so, applications will get so complicated they'll be off-loaded onto the client machines again, but for right now, let's see how COM can be used in a three- or two-tiered application.

The Evolution of COM

With what we've seen so far, the only way for COM to handle any of these configurations would be as an EXE server supporting every COM class required for an application's guts (business logic). Please see Figure 4.3.

One of the problems with this configuration, however, is that it's inflexible. Any time you want to change a single COM class, the entire EXE server has to be rebuilt and reloaded — which was to be expected years ago, but doesn't quite live up to the intentions of COM where objects just fly around a server. And maybe a lot of your application's logic is seldom, if ever, used.

Both of these problems could be solved if you could use several DLL servers instead of just one massive EXE server. However, as mentioned before, DLL servers must run inside an application — you can't load a DLL server on another machine.

What you could do, however, is redesign your COM EXE server. Because it is an application on the server, it could potentially load COM DLL servers into its process space. The only COM classes the EXE server would then support would be for communications between your client and these COM DLL servers. Please see Figure 4.4.

Figure 4.3   COM in a Tiered Environment

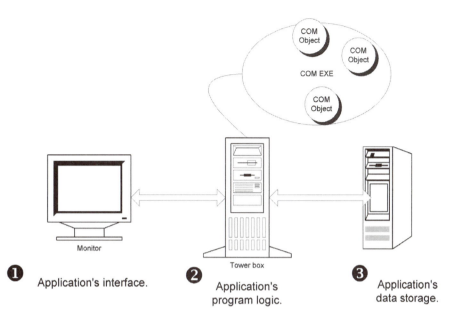

Monitor

Tower box

1 Application's interface.

2 Application's program logic.

3 Application's data storage.

Figure 4.4   COM EXE Server Using COM DLL Servers

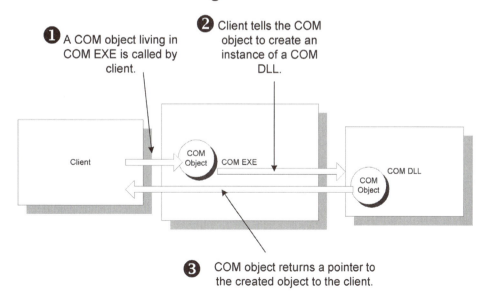

1 A COM object living in COM EXE is called by client.

2 Client tells the COM object to create an instance of a COM DLL.

3 COM object returns a pointer to the created object to the client.

But suddenly, you're a systems programmer again, figuring out how to signal to the EXE server what COM DLL you want to load and what method you want to call. Which is a shame because the work you would do on such a COM EXE server would be standard to anyone's application, so if someone has already written it, you could probably use it.

Well, they have and you can.

DLL Surrogates

A DLL Surrogate is a special EXE application that can load COM DLLs on behalf of a client — on the same machine or another machine. If you were to write a DLL Surrogate application yourself, you would have to implement the ISurrogate class and call CoRegisterSurrogate() to tell COM you're a surrogate. But more than likely, you can just use the DLL Surrogate supplied by Windows, DLLHOST.EXE.

To set your client up to use DLLHOST.EXE, you use OLE/COM Object Viewer again. This time you configure the Class ID that identifies the DLL as if it were a remote EXE. Please see Figure 4.5.

Figure 4.5 Configuring the Client for an Out-of-Process DLL

❶ Register the COM DLL as usual using regsvr32.

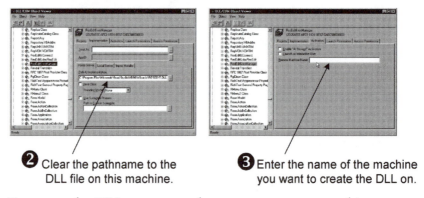

❷ Clear the pathname to the DLL file on this machine.

❸ Enter the name of the machine you want to create the DLL on.

To set up the DLL server on the same or remote machine, you set the DLL up as usual using the OLE/COM Object Viewer, but then you also check the "Use Surrogate Process" box and locate the DLLHOST.EXE application on the machine. Please see Figure 4.6.

When the client goes to create the DLL, COM redirects the call to the DLLHOST.EXE on the same machine or some remote machine and passes it the Class ID of the DLL the client wants to create.

Figure 4.6 Configuring the DLL Surrogate for an Out-of-Process DLL

❶ Register the COM DLL as usual using regsvr32.

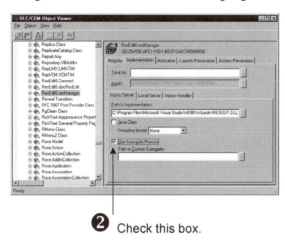

❷ Check this box.

And voilà, your DLLs are running out-of-process from your client. But beware. Now that your DLLs are out-of-process, you must make sure to either only use argument types that the COM DLL can transmit between client and server (see Chapter 2) or create and register your own proxy/stub DLL. Otherwise, your method calls won't get all the arguments it asked for.

On the other hand, a couple of interesting wrinkles pop up when using DLLHOST.EXE. First of all, your COM DLL servers can themselves create other DLL servers that will run inside of DLLHOST.EXE and therefore don't need a proxy/stub, etc. Another interesting wrinkle occurs with multitasking. As mentioned in Chapter 3, when using a COM EXE server, your multitasking choices were rather limited. But now, courtesy of DLLHOST, you suddenly get all of those choices back for your DLLs. Of course, this phenomenon is limited to the interaction between DLLs in DLLHOST.

Note: With COM objects calling other COM objects, the normal client/server terminology starts to get vague. After all, one COM object can be another object's server while at the same time be someone else's client. Therefore, the "base" client is the first client to start the proceedings. And the entire architecture is considered "multi-tiered".

Writing Your Own DLL Surrogate

What would be the advantages of writing your own DLL Surrogate? The DLLHOST application provided by Microsoft loads COM DLLs and returns their pointers back to the client. But if you had that control, what more could you do with it? How about these possibilities:

- Your DLLHOST could pool COM objects. Before any requests even come in, it could create a bunch of objects and hand them out as needed.

- To save on space, your DLLHOST could delay actually creating a COM object until the client calls one of its methods. Or it could use one of the objects from the pool.

- If you were really clever, you could set up a network of DLLHOSTs that could deflect the creation of a COM object to the least busy system in the network.

- While your DLLHOST is creating COM objects ahead of time, it could also open up a bunch of database connections and hand those out as the COM objects needed them.

- And as long as you're adding database enhancements, why not take over the work of keeping track of database transactions so that all a COM object had to do was call a method to commit or rollback a transaction. You could even provide support for transactions that span several COM objects and databases.

And by putting all of this into your version of DLLHOST, you've added functionality to all of your COM classes without adding one additional line of code.

But before you run down to the patent office, you have to know that COM already does all of this. In Windows NT, the Microsoft Transaction Server (MTS) replaces DLLHOST and provides most of this functionality.

In Windows 2000, COM+ and the operating system provide the rest.

Microsoft Transaction Server (MTS)

The Microsoft Transaction Server is an application that Microsoft originally sold as a separate package but now comes "free" on the Windows NT Options CD. As its name implies, MTS only gives you the database functionality mentioned above: MTS opens a pool of database connections for its objects and also keeps track of all of the transactions that a COM object, and the COM objects it calls, performs. Even if any one of those COM objects fail a transaction, MTS will rollback every transaction those objects performed. Please see Figure 4.7.

Figure 4.7 MTS and Transactions

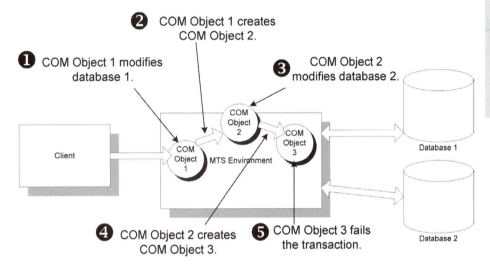

❷ COM Object 1 creates COM Object 2.

❶ COM Object 1 modifies database 1.

❸ COM Object 2 modifies database 2.

Client

COM Object 2

COM Object 1

MTS Environment

COM Object 3

Database 1

Database 2

❹ COM Object 2 creates COM Object 3.

❺ COM Object 3 fails the transaction.

❻ MTS rollbacks everything Object 1 and 2 did.

But MTS won't work with just any COM object. First of all, it obviously must reside in a COM DLL server because, at its heart, it's just a DLL Surrogate application.

To take full advantage of MTS, the object must also be written with a single database transaction in mind. Because MTS will be committing or rolling back *everything* its COM object does, it won't do for that COM object to sit there for hours in a loop handling client requests. Instead, the object should be written to handle a single request and the client should create a new COM object for each request. Please see Figure 4.8 (page 84).

Stateless Programming

Because an object should be written for a single transaction, it also doesn't make sense for it to have any long-term member variables or properties. After all, whatever is stored in the object will be gone soon anyway. Instead, the COM object should make a point of saving long-term values either in a database or some other data store. COM and other technologies call this "stateless" programming. Ironically, classes were originally developed to encapsulate data with its functionality. With MTS, there's no data, just functions.

Figure 4.8 COM Object Behavior in MTS

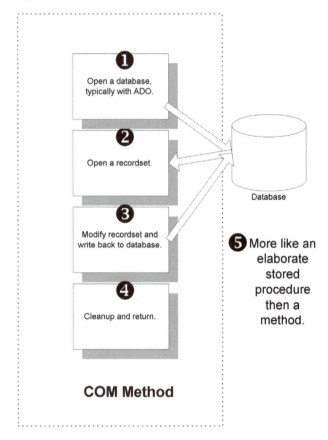

IObjectContext

MTS also needs a confederate object in your COM object to operate. This object, created from IObjectContext, is what keeps track of your object's transactions and is actually created first. Your object actually becomes the aggregate of IObjectContext (i.e., your COM class is the "derived" class and IObjectContext is the "base" class). Your object might not even exist for the entire length of a transaction.

You can also implement IObjectControl in your COM class. IObject-Context calls the functions in IObjectControl class when your object is pooled as mentioned above. Although the methods of `IObjectControl` class have no effect in MTS, they will in COM+ which supports object pooling.

Thread Safety

MTS also keeps total control over the threads running in it. You can't create your own thread from an MTS object. Nor should you want to because every object that MTS creates gets executed in its very own thread to start with.

To prevent its objects from stepping on each other's shared data, MTS relies on COM's standard threading models as shown in the tables in Chapter 3. When writing an MTS COM class using ATL, you aren't even presented with a choice for a threading model — "Both" is picked for you. That means that if this MTS object then creates an "Apartment" or "Both" object, it will run in the same thread; a "Free" will get its very own thread but with synchronization; and a "Single" will wind up running with the MTS process itself, again with synchronization.

Writing Your MTS Compliant COM Server

The Active Template Library right now is really the only practical way to write a COM server that will work with MTS. With its wizards, you can automatically add the object support you need, and ATL creates small, fast objects which are ideal for this application. When creating a COM DLL project using the ATL COM AppWizard, just make sure to check the "Supports MTS" option. And when using the ATL Object Wizard to create the object, select the "MS Transaction Server Component".

When writing the actual class:

- Avoid adding member variables to the class — they won't last for long anyway.
- Write the class to get in and out as fast as possible.
- Don't create any threads from your object.
- If your only shared resource is a database, forget about critical sections. The API will protect the database for you.

To tell MTS whether or not to commit the transactions your object(s) perform, you must call one of two methods on your object's `IObjectContext` class. If you want to commit, you call:

```
m_spObjectContext->SetComplete();
```

where `m_spObjectContext` is a smart pointer that ATL creates for you that points to the `IObjectContext` class. To rollback, you use:

```
m_spObjectContext->SetAbort();
```

If this object itself has called other COM objects, they too can either call `SetAbort()` or `SetComplete()`. All it takes is one object to call `SetAbort()` and the entire transaction is voided, even if `SetComplete()` is called later. Therefore, once it's called, you should return to the original client immediately.

You should also check to make sure `m_spObjectContext` isn't null before calling it. Although `m_spObjectContext` will never be null when running under MTS, it will be when you debug the application in the Studio.

For a complete example of writing an MTS friendly object, please refer to Example 36 (page 276).

Registering Your MTS Object

Registering a COM server for MTS is different than other COM servers. Instead of using the Regsvr32 application, you must use the Microsoft Management Console that comes with MTS. The MMC presents you with a Windows-Explorer-like interface that displays the MTS objects on your system grouped by what it calls "packages".

You can create your own package — giving it several security and activation options. You have a choice between creating a "Library" package and a "Server" package. If you want any other package to use the COM classes you add to your package, you must create a Library package. Otherwise, create a sender package. In MTS, your COM DLL servers are referred to as "components".

The MMC is also where to tell MTS whether or not you want your COM class to keep track of transactions for your object. You can even watch the "X" on the ball roll when the object has been instantiated to know it's working. For a more detailed example of registering your object, please refer to Example 37 (page 282).

When version 3.0 of MTS came out, Microsoft decided to call it COM+.

What is COM+?

Contrary to what the name implies, COM+ is not an enhancement to all of COM — although at one point, there was hope that it would be. Early presentations on COM+ showed that its developers were trying to make COM just as easy to use as any non-COM class by completely hiding its inner workings. Even developers writing early-binding COM servers would simply declare a class differently. There would be no IDL file and COM classes would support traditional inheritance.

But a year later, each month seemed to be marked by what else COM+ wouldn't be doing. Today, COM+ is more of an enhancement to MTS than any change to COM. But what they did to MTS more then compensates for what they didn't do. In fact, if you do a lot of database client/server, you will probably rarely use regular COM again.

Note: COM+ might still eventually evolve into all of its dreamed functionality, but by then, it'll probably be called COM++ or Mary's Object Emporium.

So what did they do? First, they integrated MTS into the operating system starting with Windows 2000. You can't just install COM+ by itself on a system, but by being part of the operating system, the surrogate is made much more powerful and reliable.

Then they added all of the features we supposed a really smart DLL surrogate could support plus a few more:

- **Just-in-Time Activation,** where COM+ doesn't create an object until the client actually calls one of its methods.
- **Object Pooling,** where COM+ creates a bunch of objects ahead of time.
- **Load Balancing,** where COM+ redirects a method call to the least busy machine.
- **A New Threading Model,** where COM+ provides a faster threading model that still allows it to protect your COM object.
- **Event Service,** where COM+ provides a much more robust solution to having a server callback its client (in other words, a better connection point and sink).
- **Queued Components,** where COM+ allows a client to use a COM server when it isn't even connected to the server's computer.
- **Finer-grained Security,** where COM+ allows you to specify security at the method level.

Although all of the other features we discussed in the first three chapters of this book are still available in Windows 2000, the features listed here are only available to your COM object if you configure it to use COM+. And because this functionality is being implemented in COM+, it involves little or no changes to your COM class. You write your COM object as before using ATL. But now, there are a few more options you can set when registering the object which are supported by COM+ and you can make these choices through the Component Services snap-in, the successor to MTS's Microsoft Management Console.

Just-in-Time Activation

To save on space in the server machine, COM+ can be optionally set so that it doesn't actually create an instance of your COM object for a client until the client calls one of its methods. In other words, the client asks COM+ to create an object to which COM+ replies, "I got your object right here!"

If nothing else forced you to write stateless code, just-in-time (JIT) activation does. You can store fancy data in your COM class's member variables all you want — they'll just be destroyed the minute your method returns. Please see Figure 4.9.

Figure 4.9   Just-in-Time Activation

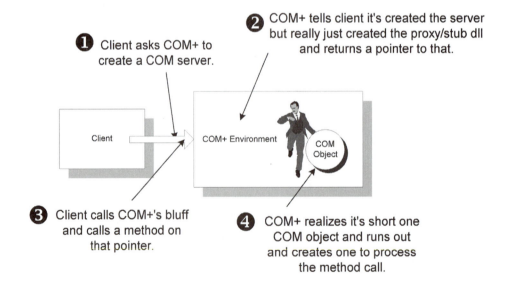

❶ Client asks COM+ to create a COM server.

❷ COM+ tells client it's created the server but really just created the proxy/stub dll and returns a pointer to that.

Client

COM+ Environment

COM Object

❸ Client calls COM+'s bluff and calls a method on that pointer.

❹ COM+ realizes it's short one COM object and runs out and creates one to process the method call.

Object Pooling

As you can image, just-in-time activation may save on memory, but it uses up precious CPU cycles in the process of creating and destroying COM objects. As a remedy for frequently-called objects, you can ask COM+ to create a few objects when the system first comes up. Then, when a client calls a method, COM+ tries to locate that object in the pool and passes the request to a match. Please see Figure 4.10.

Figure 4.10   Object Pooling

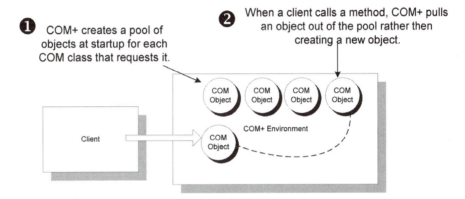

❶ COM+ creates a pool of objects at startup for each COM class that requests it.

❷ When a client calls a method, COM+ pulls an object out of the pool rather then creating a new object.

❸ If COM+ runs out of objects, it starts to create more objects but doesn't exceed a maximum value specified by the object. After that, the client has to wait.

Unlike JIT, which requires no programming on your part, you need to set up your object to support object pooling. You start by implementing the `IObjectControl` mentioned above in your COM class. Just select the appropriate option when using the ATL Object Wizard and three methods will be added to your COM class:

- `CanBePooled()` which you implement by returning `TRUE`;
- `Activate()` which you implement as if it were your COM class's constructor, initializing member variables;
- `Deactivate()` which you implement as if it were your COM class's destructor.

You also must enable object pooling for your COM class when you register it with COM+ using the Component Services snap-in. For a more specific example, please see Example 38 (page 287).

Load Balancing

Load balancing is another service you get from COM+ without changing anything in your COM class. This time when a client calls a method, it calls a "router" machine. The router machine itself can handle the call, but it also has a response-time tracker service which, as the name implies, keeps track of which machine is the least busy and can reroute the request to that machine. All of the machines eligible to handle the call are considered to be in an "Application Cluster". Please see Figure 4.11.

Figure 4.11 Load Balancing

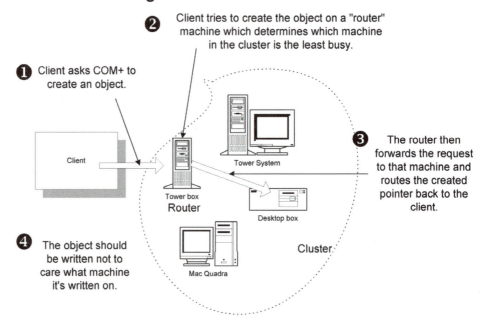

To use load balancing, you must first configure what machines are eligible to play (the cluster). To do this, you go into the router machine's properties and add the machine names to the CLBS tab.

Next, you register your COM DLL with COM+ on every machine in the cluster, making sure to also check the "Component supports dynamic load balancing" option for each. Checking this option just tells COM+ your object was written to support load balancing.

What do you do special to support load balancing? You don't use anything in your code that depended on the location of the machine that the object will be running on. In other words, you don't use a complete file path like "C:\test" and you don't even use the system time (it could be off by a minute).

Note: Load balancing is only supported by Window 2000 Servers.

A New Threading Model

Also new, but only available to COM servers running in COM+, is a new threading model. If you recall from Chapter 3, you give your object a threading model to help COM determine whether or not to provide thread safety for your object. When COM did provide safety, it was by using a message queue to synchronize access to all of the objects in a thread, thereby synchronizing their access to some shared resource — a shotgun approach to thread safety. On the other hand, you could tell COM not to provide any safety and manually add critical sections yourself to your objects to synchronize their access to a shared resource.

With COM+, you are given the option of using the "Neutral" threading model. COM automatically protects a neutral threaded object, but instead of using a message queue, it uses something more akin to your critical section — except this critical section protects the entire object. And even more than that, you can extend this critical section around any object this object calls.

How does COM+ do it? Well, it can't use a regular critical section. With normal critical sections, you call a system API function to create and lock a "flag" on your system that other threads must wait to be unlocked before they can access a section of code. COM+ can't depend on the same API call itself because the objects your object calls might span several machines and whatever flag it set up on your machine would be meaningless.

Instead, COM+ does something similar to what it does with a transaction — it passes the lock around to whatever object gets called. But, as with critical sections, when another thread tries to access a "locked" object, it has to wait until the object is done before it can proceed.

What happens though if Object A calls Object B and then Object B calls Object A? Without any further consideration, Object B would wait until Object A was finished (which won't happen because Object A is waiting on Object B.) To prevent this lockup, Object A doesn't lockout Object B because the lock is smart enough to know that B is part of the same transaction or activity.

In COM+ terminology, this is called "activity-based synchronization", where an object in any thread can participate in one sequential series of object creations and calls, but they're synchronized by this "critical section" lock that travels around with them just like a transaction. Please see Figure 4.12.

Figure 4.12 Activity-Based Synchronization

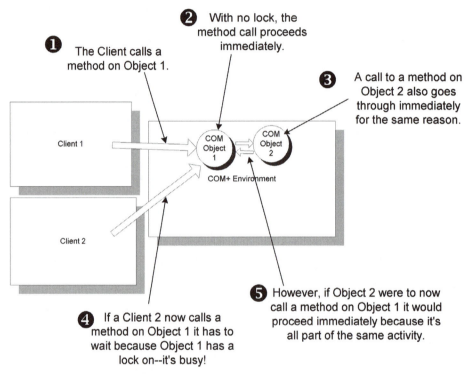

Not only do you need to give your object a "neutral" threading model, you also need to use the Component Services snap-in to configure your object to require this traveling synchronization lock. For more specifics, please refer to Example 38 (page 287).

Note: Other threading models can also be configured to support this traveling synchronization lock. However, because these objects are already getting their thread safety from other means, it wouldn't make sense to use this traveling lock as well — unless a neutral object will at some time be created along the line. All new COM+ development should use this new neutral model to provide thread security.

Event Service

In Chapter 2, we reviewed two options for a COM server to call a client back when an event occurred. These included adding a connection point to the server and a sink to the client, or using ActiveX's standard event firing through a Windows message. COM+ has a new service that allows a server to talk to a client, but it's not what you might think.

First of all, this service is again only available to objects running under COM+. You start by creating a dummy "event" class that defines the class and methods a COM server will be calling when an event occurs — not unlike the sink interface we had to define in Chapter 2 to tell the COM server with the connection point what method to call. This time however, you register this class with COM+ as an event class.

The dummy class you created doesn't actually implement anything — its method calls simply return to the caller. But to write the actual COM+ client objects, you would use this dummy class's interface and actually implement it like any other COM+ class. You then register these client classes with COM+, only this time you tell COM+ that this object should be created and called when the previously registered event class is created and called.

And that's it. For a COM server to activate these client objects when an event occurs ("fire an event") now, it just creates the event class and calls its methods. Please see Figure 4.13 for an overview or Example 39 (page 294) for more details.

Figure 4.13 COM+ Event Service

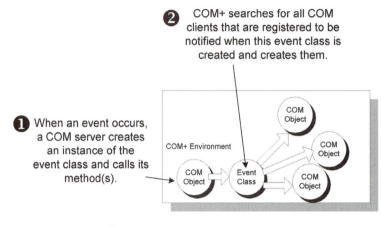

❷ COM+ searches for all COM clients that are registered to be notified when this event class is created and creates them.

❶ When an event occurs, a COM server creates an instance of the event class and calls its method(s).

COM+ Environment

COM Object

Event Class

COM Object

COM Object

COM Object

COM Object

❸ But you have to be a COM+ object to participate.

The COM+ team celebrated the creation of this service with two new terms. The COM+ server is now called a "publisher" and the clients are called "subscribers". And perhaps it's just as well they did rename these players because the application of this technology isn't quite targeted to the same set of problems — a sink and connection points. Instead of solving the problem of informing some external client when an event occurs (such as an administrator application), the clients in this case are COM+ objects too. Therefore, they can only do what COM+ objects can do: handle a transaction and return as quickly as possible.

So what this technology really provides is a way for one COM+ object to create and call a bunch of other COM+ objects with just one call.

Queued Components

For all of the last three and a half chapters, we assumed that when a client wanted to create and call the methods of a COM server, that that server was available. But what about the case of a client on a laptop computer that wants to talk to a COM server on some central computer? Do we forbid the client application from even running until the laptop is on the network? With COM+, you don't have to. Instead, you can use what's called a *queued component*.

Once again, you can only use queued components with COM+. This time it's because COM+ provides three COM servers of its own that can record and playback method calls. The client uses the record server (QC.Recorder) to record its method calls and the server uses the playback object (QC.Player). These two COM+ objects in the middle talk to each other — not with COM this time, but with the Microsoft Message Queue (MSMQ), which is a relatively new API (12/97) available to any application that wants to queue messages to an application on the same machine or another. I highly recommend it for your queuing needs.

To take advantage of COM+'s queued components, you start by creating your COM server as always except this time you don't specify any method arguments as returning data back (i.e., as being an [out] or [in,out]). After all, it wouldn't make sense for the client to hang around after it's made a call to see what the server has to return.

Next, when you register the server with the Component Services snap-in, you tell it that the object in this server is eligible to process messages straight from the MSMQ and that its interface is queued. And finally, you also register the server on the client machine telling it what machine the server is on — as you would with any COM server using Regsvr32 and the OLE/COM Object Viewer.

The client at this point could create an instance of the server and use its methods as always but the communication would not be queued. Instead, the client must create an instance of the COM+ server that records method calls (QC.Recorder). When the client creates the recording object, it lets it know for which real COM server its recording method calls.

The recording object then exposes those methods to the client as if it were the real COM server, and the client proceeds as if it were talking to the real server. Creating and talking to the recording server might look something like this:

```
LPUNKNOWN lpUnknown=NULL;
::CoGetObject("queue:/new:Server.Wzd",NULL,IWzd,&lpUnknown);
IWzd *pWzd=(IWzd*)lpUnknown;
pWzd->Method1(1234);
```

At the point where the client finally releases the recording object, the recording object formats a message and sticks it into an MSMQ queue. MSMQ takes over and attempts to send the message to a queue on the designated machine. How does it know what machine to send it to? It gets it from the machine address you specified when registering the COM server on the client machine.

If the server machine is up and running, the message flies across to it. Otherwise, MSMQ holds onto the message until the target machine does appear on the network.

On the server machine, yet another COM+ server object is listening to MSMQ to see if any messages have come in for the objects in its package. When one does come in, this object (QC.ListenerHelper) creates the playback object (QC.Player) to create and playback the method calls.

For an overview of this process, please see Figure 4.14. For more details, please refer to Example 40 (page 296).

Figure 4.14   Queued Components

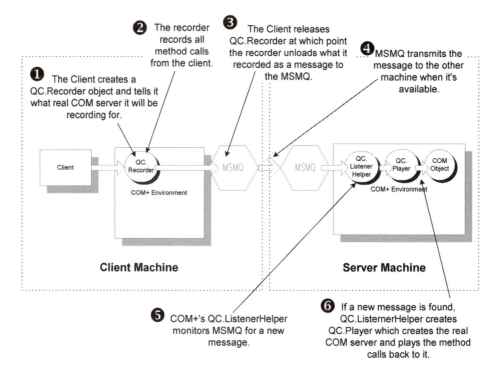

Queued components even work with transactions. In other words, if a queued component is one of the components called during a transaction, it will abort the transaction if it can't successfully place the message in an MSMQ queue. For obvious reasons, however, it can't also abort the transaction if the server never gets the message or can't process it.

On the positive side, MSMQ will make a Herculean effort to deliver the message, even storing it in a dead letter queue if it never succeeds. But if the laptop falls off a bridge, what can you do?

Security at the Method Level

We covered security for most COM applications in the last chapter. Before COM+, security could be provided in two ways:

- You could prevent any COM object from being created remotely on a machine,

- Or you could require that only certain users are able to create a COM object remotely.

If you use COM+, you can now also apply security at the method level. In other words, you can require that just about anybody can create an object remotely and use most of its methods, but only certain User IDs are allowed to access the dangerous methods.

To specify security at the method level, you once again use the Component Services snap-in. And as with object level security, the list of users that can be authorized comes right out of Windows 2000's User Manager. But to avoid assigning the same list of individuals to each method call, COM+ allows you to group individuals into what are called "roles" (as in acting roles or your role in life). In terms of security, someone in the role of police chief will be able to access dangerous methods while the rookie won't.

But what happens once you've create an object and that object wants to create another object — will it have to go through security again? Not if it uses another object in the same COM+ package. On the other hand, if it does leave the package, it will need to go through security again. For this reason, you will need to configure a User ID and password for it when it does.

The Component Services snap-in gives you two choices for User IDs. The first allows the object to use the same User ID as the person currently logged into the server machine. This is also great for debugging because any messages appear to the user. On the negative side, the object won't get anywhere if no user is logged on. So for the release version of your application, you should use Component Services snap-in second option of specifying the exact User ID and password to use.

For more details, please see Example 38 (page 287).

Other Minor Changes

Besides these major enhancements to MTS, you should also be aware of two minor enhancements. First of all, the `IObjectContext` class offers you two new methods you can call. If you remember from earlier, `IObjectContext` already provides you with the methods: `SetCompletet()` and `SetAbort()` which allows your COM server to tell COM+ to commit or rollback a transaction. The two new methods are called `EnableCommit()` and `DisableCommit()` which allow you greater control over a transaction.

When a COM+ object returns to the original client, COM+ immediately deactivates the object — that means it either destroys the object or returns it to the object pool. However, if you call `DisableCommit()` before you return to the client, COM+ won't do anything to the object. Therefore, by using `DisableCommit()`, you would be able to keep calling methods on the same object until you call `EnableCommit()` and return. What's the bottom line? Your object can maintain member variables over several calls — in other words, you would have an object that maintains its state while all the other objects around it are stateless.

Another minor change is yet another name change. COM+ developers in their wisdom have decided that a package in MTS is an "application" in COM+. Yet another chapter in COM's campaign to confuse its users.

Attribute Programming

COM+ represents a shift in programming philosophy. Although application programmers weren't suppose to know much about system programming, as the requirements for applications evolved, they required more and more systems programming to meet those requirements. Application programmers had to be aware of thread safety and client/server protocols and how best to scale their application up to handle thousands of users.

With COM+, application programmers no longer need to worry about adding code to their application for thread safety or protocols or scalability or security. Instead, they get this functionality from COM+, just by activating the right setting.

In COM+ terminology, setting options to get thread safety, security, etc. — instead of writing code — is called "attribute programming".

The Component Catalog

And speaking of all these new attributes, the system registry where they were stored in Windows NT and 9x has finally burst its seams. So all new settings that relate to COM+ have found a new home in a file called the Component Catalog. COM objects that don't live in the COM+ environment will still continue to live in the system registry for years to come.

COM+ vs. EJB

Before I finish this chapter, I thought I should briefly discuss COM+'s main competitor, just to give you an idea of everything that's available. Just about the same time the original COM was rolling out, a competing standard was forming called CORBA. Unlike COM, CORBA was to allow the same functionality of calling any object from any client *and* you would also be able to call from any platform — not just Microsoft platforms. Unfortunately, the flexibility that CORBA now allowed you also became its downfall — a gaggle of lesser vendors couldn't make it a widely-accepted standard. All Microsoft had to do was put COM in their operating system for free and suddenly everybody had it.

But out of the ashes of CORBA has risen yet another challenger to COM/COM+ called Enterprise Java Beans or EJB. EJB hopes to deliver the same functionality as COM+ (there's a liberal sharing of ideas) but again on any platform.

Note: Not everyone would consider CORBA dead but Sun did retire their CORBA product (NEO) when they brought out EJB. As with all technologies, CORBA will be around until everything that uses it is retired. More like Betamax then eight-track cassettes.

This time, however, EJB has the backing of another powerhouse vendor, Sun Micro, the same people that brought you Java. As you might expect, one of the downsides to EJB is that objects can only be written in Java. Another is the lack of features you might expect with any new product. But right now, Sun Micro is a bigger name on Unix machines and EJB will be a force to reckon with there.

Summary

In this chapter, we saw COM+ evolve from the simple DLLHOST application that could host a COM DLL on another machine into a new programming philosophy with COM+. We saw MTS develop to take over the chores of keeping track of transactions over several objects or systems. And we saw MTS evolve into COM+, allowing you to pool your objects or divide their labor over several machines or secure them against intrusion, all without writing one line of code ourselves.

COM+ and EJB represent the next step in the evolution of software programming. First there was **spaghetti programming** where your application was just page after page of subroutines and goto's. Then there was **object-oriented programming** where your application was split up into classes that could be independently tested and reused. And now there is **attribute programming** where programming is more a matter of setting the right option and writing the actual object is almost an afterthought. And as with past evolutionary steps, it will take time to get used to not having to worry about threading, security, etc. But to paraphrase one popular COM+ author, "Don't worry, be COM". (Get it? COM sounds like "calm". It also has that "Don't worry, be happy" kind of thing going for it.)

After all, COM+ is just the next logical step while we're all just waiting around for computers to be fast and smart enough to know what we want without us having to program them.

This ends the basics portion of this book. The remaining chapters take an example-centric approach to COM, presenting common client/server problems and their COM solutions. I tried to cover all the various incarnations of COM because even though COM+ represents an exciting new way to program, it really doesn't offer the whole package yet.

COM Examples

The examples in this book concentrate on getting you writing COM objects for a majority of applications as quickly as possible. We start with creating someone else's COM object, progress to writing your own using ATL, MFC, Visual Basic and Visual J++, review the argument types you can use to talk to these objects, and graduate with writing your own COM+ objects. Towards the end, we also review one of Microsoft's prewritten COM object libraries called Active Data Objects (ADO) to access data sources (ex: databases).

The topics covered in this section by chapter include:

Creating and Accessing COM Objects

Examples in Chapter 5 include creating a COM object using C++ with the COM API directly or with smart pointers where the objects support early or late binding. Creating objects with Visual Basic and Visual J++ are also included, as well as creating ActiveX controls.

Writing COM Servers with MFC

In Chapter 6 are examples of writing your own COM server using the MFC class library. We look at writing DLL or EXE servers that support early and late binding, aggregation, licensing, connection points, and singletons. There's also a quick review of writing an ActiveX control.

Writing COM Servers with ATL

Chapter 7 shows how to write most of the same MFC COM servers that usually use ATL more easily. Additional examples here show how to write a COM NT Service, tear-off interfaces, and an extended interface.

Writing COM Servers with VB and VJ++

Although writing a server with Visual Basic and Visual J++ is easier then with ATL or MFC, you have less options as well. Chapter 8 has examples of writing COM servers using these two languages and also how to provide a sink for an ATL COM object with a connection point.

COM Communications

Chapter 9 has several examples of how to use COM to automatically send data between your client and COM server. Included is an example of all the data types that COM supports as well as Visual Basic and Visual J++.

COM+ Examples

In Chapter 10, we show some of the things you can do with the latest enhancements to COM. Included is an example of writing a COM+ object, using the ability for a client to queue its call to a COM server until the COM server is available, and using the event server provided by Windows 2000.

Accessing Database Objects

The three examples in Chapter 11 show how to use Microsoft's ADO COM objects to access a database. The first example puts ADO through its paces, creating and accessing data elements using C++, while the last two chapters do the exact same thing for VB and VJ++. Sort of an ADO Rosetta stone.

Potpourri

The examples in this last chapter (12) are an eclectic bunch — from how to add licensing support to your COM server, to processing COM errors, to using the free threaded marshaller to prevent COM from ever trying to make your COM object thread safe.

Chapter 5

Creating and Accessing COM Objects

Before we get into all the complications involved with writing a COM class ourselves, I thought a review of how easy it is to use a COM object would help prove to you that it's all worth while. We will look at several of the most common permutations of creating a COM object, from using the API directly to being totally oblivious to the whole procedure in Visual Basic. And from creating totally custom COM classes to creating ActiveX controls.

The examples in this chapter include:

Example 1   Creating a COM Object Using C++ and the COM API
where we roll up our sleeves and revert to the API level to create a COM object.

Example 2   Creating a COM Object Using C++ and Smart Pointers
where we see how object-oriented programming simplifies creating COM object.

Example 3   Creating a COM Object Using MFC and Late Binding where we review how to create a COM object that only supports late binding (aka, Automation) using MFC classes.

Example 4   Creating a COM Object Using Smart Pointers and Late Binding where we review how to create a COM object that only supports late binding using smart pointers.

Example 5   Creating an ActiveX Control Using MFC where we discover how to add an ActiveX control to our MFC project.

Example 6   Creating an ActiveX Control from Visual Basic where we discover how easy it is to add an ActiveX control to this classic scenario.

Example 7   Creating a COM Object from Visual Basic where we review the different VB syntaxes you can use to create a COM object.

Example 8   Creating a COM Object from Visual J++ where we do the same for VJ++.

Example 1   Creating a COM Object Using C++ and the COM API

Objective

You would like to create and access a COM object by calling the COM API directly from C++.

Strategy

We will look at three common ways to create a COM object using the COM API directly. The API functions include, in order of importance: `::CoCreateInstance()`, `CoCreateInstanceEx()`, and `::CoGetObject()`. The last function is usually only of importance as a diagnostic tool. Both `::CoCreateInstance()` functions call `::CoGetObject()` internally and by calling it ourselves, we can help troubleshoot an object creation problem.

This example is included for the rare case of when you need to create a COM class directly with the API. For a simpler approach, please refer to the next example.

Example 1   Creating a COM Object Using C++ and the COM API   **105**

Steps

Include the Identifiers for the COM Server in Your Project

The COM class you want to create is identified in the system by guids, which are 128 bit arrays containing a set of numbers theoretically unique throughout the universe. Both the DLL or EXE file the class lives in and the class itself have their very own ids. The identifier for the DLL/EXE file is called a Class ID and the id for the COM class itself is called an Interface ID (confused yet?). Because the ids are arrays, they must be included in your project as data items as opposed to simple defines or constants. In other words, you have to create a static global array per id in one source file, and if these ids are required in another file they must be declared externally there.

1. Create a guids.h file.

2. Add the class and interface ids for all of the COM classes that you'll be creating in your project. If the COM class was created using ATL, just cut and paste the ids you'll find in the xxx_i.c file created by that project. For MFC COM classes, locate and copy the files from the xxx_i.c file located in the separate Interface Server project (see Example 9, page 143). This, unfortunately, only contains the interface ids. You will need to dig into the .cpp file in the MFC project for the IMPLEMENT_OLECREATE macro to find the class id. If there are multiple COM classes in this MFC project, you still only need one class id. The final guids.h file should look something like this:

```
#if !defined guids_h
#define guids_h

const IID CLSID_IWzdSrv = { 0x4487d432, 0xa6ff, 0x11d3, 0xa3, 0x98,
    0x0, 0xc0, 0x4f, 0x57, 0xe, 0x2c };
const IID IID_IWzd =
    {0xC177116E,0x9AAA,0x11D3,{0x80,0x5D,0x00,0x00,0x00,0x00,0x00,0x00}};

#endif
```

Initialize the COM DLL

Unlike other Win32 APIs, you need to tell the COM API — specifically OLE32.DLL — that you're going to be using it today. There is one of two calls

II   5

you can make to do this as shown in the next two steps. You must make this call not only when your application first starts, preferably in the `InitInstance()` of an MFC application, but also at the start of any thread your application creates.

1. Your application can call:

```
::CoCreateInstance(
        NULL        //reserved
);
```

OR

2. Your application can call:

```
::CoInitializeEx(
    NULL,                       //always NULL
    COINIT_APARTMENTTHREADED  //or COINIT_MULTITHREADED
    );
```

You also need to add _WIN32_DCOM to your project settings under "Preprocessor definitions" in order to get the prototype definition for `::CoInitializeEx()` included in your compile.

The second API function allows for better control of your COM objects in a multitasking environment. Please refer to Chapter 3 for more on how a COM object behaves in a multitasking environment.

Include the COM Definition Files

1. Include the `guids.h` file as well as the COM class's `.h` file (also generated by MIDL) and an MFC `.h` file:

```
#include "IServer\IWzd.h"
#include "guids.h"
#include <objbase.h>
```

Remember — only include the `guids.h` file once. The variables it defines are global.

Example 1   Creating a COM Object Using C++ and the COM API   **107**

Create the COM Object

1. To create a COM object one at a time, use:

```
IWzd *iWzd=NULL;
HRESULT hr=::CoCreateInstance(
        CLSID_IWzdSrv,      // name of dll to load
        NULL,               // aggregated COM object (none)
        CLSCTX_ALL,         // most flexible option
        IID_IWzd,           // class to create and object of
        (LPVOID*) &iWzd);   // returned object pointer
if (FAILED(hr))
{
    _com_error err(hr);
    AfxMessageBox(err.ErrorMessage());
    return;
}
```

2. Use the object like any C++ class:

```
iWzd->Method1(…);
```

3. To destroy the object, use:

```
iWzd->Release();
```

4. To call what `CoCreateInstance()` calls directly (for diagnostic reasons), use:

```
// load DLL/EXE and get a pointer to its class factory
IClassFactory *pCF=NULL;
hr=::CoGetClassObject(
   CLSID_IWzdSrv,           // name of dll to load
   CLSCTX_ALL,              // most flexible
   NULL,                    // for DCOM, a COSERVERINFO
                            // structure that id's the remote server
                            // more typically set using OLEView
   IID_IClassFactory,       // the class factory interface (all
                            //COM DLL/EXE's must have this interface)
   (LPVOID*)&pCF);
if (FAILED(hr))
```

II   5

```
{
    _com_error err(hr);
    AfxMessageBox(err.ErrorMessage());
    return;
}
// ask class factory to create the class
hr = pCF->CreateInstance(
    NULL,                       // aggregated COM object (none)
    IID_IWzd,                   // class to create and object of
    (LPVOID*) &iWzd);           // returned object pointer
if (FAILED(hr))
{
    _com_error err(hr);
    AfxMessageBox(err.ErrorMessage());
    return;
}
pCF->Release();
iWzd->Release();
```

5. To create multiple COM objects at once, use:

```
// setup MULTI_QI structure with all the classes to create
MULTI_QI results[3];
memset(results,0,sizeof(results));
results[0].pIID=&IID_IWzd;  // class to create
results[1].pIID=&IID_IWzd;
results[2].pIID=&IID_IWzd;


// make the call
hr =::CoCreateInstanceEx(
            CLSID_IWzdSrv,      // name of dll to load
            NULL,               // aggregated COM object (none)
            CLSCTX_INPROC_SERVER,// use dll
            NULL,               // for DCOM, a COSERVERINFO
structure that id's the remote server
                                // more typically set using OLEView
            3,                  // number of MULTI_QI structures in
```

Example 1   Creating a COM Object Using C++ and the COM API   **109**

```
results
              (MULTI_QI*)&results); // array of MULTI_QI structures
if (FAILED(hr))
{
   _com_error err(hr);
   AfxMessageBox(err.ErrorMessage());
   return;
}
for (int i=0;i<3;i++)
{
    if (FAILED(results[i].hr))
    {
         _com_error err(hr);
         AfxMessageBox(err.ErrorMessage());
    }
}
```

6. You'll then need to get the pointers to the created objects from the MULTI_QI structure:

```
IWzd *iWzd1=(IWzd*)results[0].pItf;
IWzd *iWzd2=(IWzd*)results[1].pItf;
IWzd *iWzd3=(IWzd*)results[2].pItf;
// release all objects when done
iWzd1->Release();
iWzd2->Release();
iWzd3->Release();
```

Notes

- You'll find that almost all COM API function calls start with a ::Co... prefix for "Com Object". When writing your own COM object, you might consider naming your COM class this way (e.g., CoAccessDB) in order to differentiate COM classes from regular C++ or Java classes.

- We used CLSCTX_ALL in the API class above. The other options allow you to specify exactly what type of file the COM class lives in: DLL or EXE. However, CLSCTX_ALL gives you the flexibility of allowing someone at configuration time to determine whether to use a version of the class that lives in an EXE file or a DLL file. When given the choice, COM checks for DLLs first, then EXEs that use the exact same Class ID.

CD Notes

- When executing the project on the accompanying CD, put a breakpoint in the OnTest() button handler and step through the API calls.

Example 2   Creating a COM Object Using C++ and Smart Pointers

Objective

You would like to create and access a COM object by using the smart pointer class and the #import compiler directive.

Strategy

Unlike the direct API approach shown in the last example, smart pointers bring all of the niceties of object-oriented programming to the creation of your COM objects. Specifically, the #import compiler directive takes care of adding those messy guids to your project. And the API calls themselves are encapsulated in custom classes that the #import directive also creates. These classes not only make creating an object easier, but also helps ensure the COM class is destroyed since the Release() method is called from its destructor.

Steps

Import the COM Class Definitions

1. Import the COM class's definitions into a project source file using the #import directive. The file you include can be a DLL, EXE, TLB, or OCX file. The #import directive has several keywords you can use. Without any keywords, the classes that #import creates require you to specify a scope for every reference to the COM class (i.e., SERVERLib::IWzdPtr)

Example 2   Creating a COM Object Using C++ and Smart Pointers   **111**

where the scope name is the name of the Library statement in the IDL file:

```
#import "server\server.tlb"
```

2. When none of the values you're importing interfere with any definitions already in that source file, you can avoid having to use the scope operator by using this keyword with the #import directive:

```
#import "server\server.tlb" no_namespace
```

3. Using the following keyword keeps the compiler from adding extra error checking code around every method call you make to the imported COM object. If you want to customize the way you handle errors, use this:

```
#import "server\server.tlb" raw_interfaces_only
```

Also see Example 46 (page 333) for more on automated COM error handling.

Initialize the COM DLL

Unlike other Win32 APIs, you need to tell the COM API — specifically OLE32.DLL — that you're going to be using it today. There is one of two calls you can make to do this shown in the next two steps. You must make this call not only when your application first starts, preferably in the InitInstance() of an MFC application, but also at the start of any thread your application creates.

1. Your application can call:

```
::CoCreateInstance(
    NULL          //reserved
);
```

OR

2. Your application can call:

```
::CoInitializeEx(
    NULL,                       //always NULL
    COINIT_APARTMENTTHREADED  //or COINIT_MULTITHREADED
    );
```

You also need to add _WIN32_DCOM to your project settings under "Pre-processor definitions" in order to get the prototype definition for ::CoIni-tializeEx() included in your compile.

The second API function allows for better control of your COM objects in a multitasking environment. Please refer to Chapter 3 for more on how a COM object behaves in a multitasking environment.

Create the COM Object

1. The #import directive created a class wrapper you can now use to create your COM class:

```
IWzdPtr pPtr1(
    __uuidof(Wzd)//class id of DLL or EXE that contains class
        );
```

where the class name, IWzdPtr, is the original interface name (i.e., IWzd) with a "Ptr" suffix. Looking at this syntax another way, you want to create the COM class IWzd which is located in the Wzd DLL or EXE file.

2. To create the object using a smart pointer's method instead of in its constructor so that you can get the error code, use:

```
IWzdPtr pPtr2;
HRESULT hr=pPtr2.CreateInstance(
    __uuidof(Wzd)  //class id of DLL or EXE that contains class
            );
```

3. To call the member variables of the COM object, use this:

```
pPtr1->Method1(1234);
```

Note that even though the class is created on the stack, the pointer operator is used. Smart pointers override the pointer operator to denote you are accessing the methods of the created object. To access the methods of the smart pointer class itself (i.e., QueryInterface()), use the dot syntax (pPtr.QueryInterface()).

4. To catch COM errors, use:

```
try
{
    ...
}
```

Example 2   Creating a COM Object Using C++ and Smart Pointers   **113**

```
catch (_com_error &err)
{
    AfxMessageBox(err.ErrorMessage());
    return;
}
```

5. Unless you use the `raw_interfaces_only` attribute when importing a COM class's type library, method calls will check to make sure a object is created and will also throw a `_com_error` structure if the method returns a failure. If no error occurs, the method call will also make any `retval` argument into a returned value:

```
HRESULT hr=pPtr1->Method1(&retval);
```

II   5

becomes:

```
long retval=pPtr->Method1();
```

Please refer to Example 46 (page 333) for more on COM error handling.

Notes

- If you look in your Debug or Release directory of your client project after you've used the #import directive, you'll notice it created a .tli and a .tlh file for every file it imported. Inside you'll find the custom classes that #import created along with the guids you no longer need to worry about. Please refer to the following #Import Generated File Listings for an example.

- The scope operator name (i.e., SERVERLib) comes from the type library's name (the Library statement in the IDL file). The class id name comes from the coclass name also found in the type library or IDL file.

CD Notes

- When executing the project on the accompanying CD, put a breakpoint in the OnTest() button handler and step through the API calls.

Listings — #Import **Generated TLH File**

```
// Created by Microsoft (R) C/C++ Compiler Version 12.00.8168.0
    (6478b555).
//
// g:\creating a com object using cpp and smart
    pointers\debug\server.tlh
//
// C++ source equivalent of Win32 type library server\server.tlb
// compiler-generated file created 12/24/99 at 09:00:00 - DO NOT EDIT!


#pragma once
#pragma pack(push, 8)

#include <comdef.h>

//
// Forward references and typedefs
//

struct /* coclass */ Wzd;
struct __declspec(uuid("351495d0-aa6d-11d3-807e-000000000000"))
/* dual interface */ IWzd;

//
// Smart pointer typedef declarations
//

_COM_SMARTPTR_TYPEDEF(IWzd, __uuidof(IWzd));

//
// Type library items
//

struct __declspec(uuid("351495d1-aa6d-11d3-807e-000000000000"))
Wzd;
    // [ default ] interface IWzd
```

Example 2   Creating a COM Object Using C++ and Smart Pointers   **115**

```
struct __declspec(uuid("351495d0-aa6d-11d3-807e-000000000000"))
IWzd : IDispatch
{
    //
    // Wrapper methods for error-handling
    //

    HRESULT Method1 (
        long lArg );

    //
    // Raw methods provided by interface
    //

    virtual HRESULT __stdcall raw_Method1 (
        long lArg ) = 0;
};

//
// Wrapper method implementations
//

#include "g:\creating a com object using cpp and smart
    pointers\debug\server.tli"

#pragma pack(pop)
```

Listings — #Import **Generated TLI File**

```
// Created by Microsoft (R) C/C++ Compiler Version 12.00.8168.0
    (6478b555).
//
// g:\creating a com object using cpp and smart
    pointers\debug\server.tli
//
// Wrapper implementations for Win32 type library server\server.tlb
// compiler-generated file created 12/24/99 at 09:00:00 - DO NOT EDIT!
```

```
#pragma once

//
// interface IWzd wrapper method implementations
//

inline HRESULT IWzd::Method1 ( long lArg ) {
    HRESULT _hr = raw_Method1(lArg);
    if (FAILED(_hr)) _com_issue_errorex(_hr, this, __uuidof(this));
    return _hr;
}
```

Example 3   Creating a COM Object Using MFC and Late Binding

Objective

You would like to create and access a COM object that only supports late binding using MFC.

Strategy

In the last two examples, we created COM objects using what is called their "custom interface". You can also use those examples to create a COM object that supports both a custom interface and a late binding or automation interface (a "dual" interface). Few and far between COM objects exists that only support a late-binding interface. Those types of COM objects only allow access to its functionality through an IDispatch class implementation and represent the first types of COM objects. Please refer to Chapter 1 for more on types of COM objects.

We will be using the ClassWizard to create a C++ class that will wrap a COM object that supports late binding. We will also look at an MFC class that helps handler errors.

Example 3   Creating a COM Object Using MFC and Late Binding   **117**

Steps

Initialize the COM DLL

Unlike other Win32 APIs, you need to tell the COM API — specifically `OLE32.DLL` — that you're going to be using it today. There is one of two calls you can make to do this shown in the next two steps. You must make this call not only when your application first starts, preferably in the `InitInstance()` of an MFC application, but also at the start of any thread your application creates.

1. Your application can call:

```
::CoCreateInstance(
      NULL         //reserved
);
```

OR

2. Your application can call:

```
::CoInitializeEx(
      NULL,                      //always NULL
      COINIT_APARTMENTTHREADED //or COINIT_MULTITHREADED
      );
```

You also need to add `_WIN32_DCOM` to your project settings under "Preprocessor definitions" in order to get the prototype definition for `::CoInitializeEx()` included in your compile.

The second API function allows for better control of your COM objects in a multitasking environment. Please refer to Chapter 3 for more on how a COM object behaves in a multitasking environment.

Create a C++ Class to Encapsulate the COM Class

1. Create a helper class that will wrap this COM object by first opening the ClassWizard and clicking on the Member Variables page. Then, click on the "Add Class..." button and select the "From a type library" menu item. Then, find and select the COM object's `.tlb` file.

Create and Use a Late Bound COM Class

1. Include the `.h` file created above in your source file.

2. Create an instance of the class and use its `CreateDispatch()` method to create the actual COM object:

```
COleException *e=new COleException;
IWzd wzd;
if (!wzd.CreateDispatch(
        CLSID_IWzdSrv,            // or Program ID (ex:"Server.Wzd.6")
        e))
   throw e;
```

3. You can now access its methods as with any C++ class:

```
wzd.Method1(1234);
wzd.SetProperty1(4321);
long lVal=wzd.GetProperty1();
```

4. To catch any COM errors using MFC's `COleException` class, you can use:

```
COleException *e=new COleException;
try
{

}
catch (COleException *e)
{
    _com_error err(e->m_sc);
    AfxMessageBox(err.ErrorMessage());
    e->Delete();
}
```

Notes

- The C++ class that the ClassWizard generates simply converts your method calls into the call configuration required by the `IDispatch` class. Please see the following example in the Generated C++ Class Listings.

- As mentioned in Chapter 1 and 2, this type of interface is less efficient than a custom interface because a custom interface is a direct call to the server (when in a DLL) while a late bound call has the additional overhead required by `IDispatch` to get to the right function. When the server is in an EXE however, the overhead of interprocess communications overshadows this inefficiency.

Example 3   Creating a COM Object Using MFC and Late Binding   **119**

- When creating a COM object from Visual Basic or Visual J++, the late-binding interface is used exclusively. However, this connection can be optimized by querying the object at compile-time for its functionality as it's done here with MFC and the class generated by the ClassWizard.

CD Notes

- Build the project on the CD is debug mode. Put a breakpoint in the OnTest() button handler and step through as the object is created and called.

Listings — Generated C++ Class .H File

```
// Machine generated IDispatch wrapper class(es) created with
   ClassWizard
///////////////////////////////////////////////////////////////////
    /////////
// IWzd wrapper class

class IWzd : public COleDispatchDriver
{
public:
   IWzd() {}              // Calls COleDispatchDriver default constructor
   IWzd(LPDISPATCH pDispatch) : COleDispatchDriver(pDispatch) {}
   IWzd(const IWzd& dispatchSrc) : COleDispatchDriver(dispatchSrc) {}

// Attributes
public:
       long GetProperty1();
       void SetProperty1(long);

// Operations
public:
   void Method1(long lArg);
};
```

Listings — Generated C++ Class .CPP File

```cpp
// Machine generated IDispatch wrapper class(es) created with
   ClassWizard

#include "stdafx.h"
#include "server.h"

#ifdef _DEBUG
#define new DEBUG_NEW
#undef THIS_FILE
static char THIS_FILE[] = __FILE__;
#endif




/////////////////////////////////////////////////////////////////////
   /////////
// IWzd properties

long IWzd::GetProperty1()
{
   long result;
   GetProperty(0x1, VT_I4, (void*)&result);
   return result;
}

void IWzd::SetProperty1(long propVal)
{
   SetProperty(0x1, VT_I4, propVal);
}

/////////////////////////////////////////////////////////////////////
   /////////
// IWzd operations

void IWzd::Method1(long lArg)
```

```
{
    static BYTE parms[] =
        VTS_I4;
    InvokeHelper(0x2, DISPATCH_METHOD, VT_EMPTY, NULL, parms,
        lArg);
}
```

Example 4   Creating a COM Object Using Smart Pointers and Late Binding

Objective

You would like to create and access a COM object that only supports late binding using smart pointers.

Strategy

In the last example, we created a late-binding COM object using MFC's classes. But just as smart pointers made creating an object that supports early binding easier, so does it make creating a late-bound object. In fact, unless you looked at the .tlh and .tli files generated by the #import directive, you wouldn't know the object was any different.

Steps

Initialize the COM DLL

1. Your application must call:

```
::CoCreateInstance(
     NULL          //reserved
);
```

OR

2. Your application must call:

```
::CoInitializeEx(
    NULL,                     //always NULL
    COINIT_APARTMENTTHREADED  //or COINIT_MULTITHREADED
    );
```

You also need to add _WIN32_DCOM to your project settings under "Pre-processor definitions" in order to get the prototype definition for ::CoInitializeEx() included in your compile.

Create the COM Object

1. Import the type library or DLL or EXE file of the server using one of several formats:

```
#import "server\server.tlb"              // must specify
   SERVERLib:: namespace scope in
                                         // front of every IWzdPtr, etc.
```

OR

```
#import "server\debug\server.tlb" no_namespace // don't need to use
   a SERVERLib:: scope in
                                         // front of every

   IWzdPtr, etc.
```

OR

```
#import "server\server.tlb" raw_interfaces_only // doesn't create
   wrappers around each method
```

2. Create the COM object using the smart pointer:

```
    IWzdPtr wzd(
__uuidof(Wzd)
               //guid of DLL or EXE that contains class
               );
```

3. Call the methods of the COM object:

```
    wzd->Method1(1234);
    wzd->PutProperty1(4321);
    long lVal=wzd->GetProperty1();
//(The smart pointer takes care of converting your method calls )
   into Invoke()'s.
```

Example 5   Creating an ActiveX Control Using MFC   **123**

4. Catch any COM errors with:

```
try
{
}
catch (_com_error &err)
{

    AfxMessageBox(err.ErrorMessage());
    return;
}
```

Notes

- Check what class name to use from the .tli file that the #import directive extracts from the type library. For example, in the case of VJ++, the name extracted is actually XxxDispatchPtr, where Xxx is the class name you use inside of VJ++.

CD Notes

- Build the project on the CD is debug mode. Put a breakpoint in the OnTest() button handler and step through as the object is created and called. Notice that the debugger won't let you step through the dispatch call, so you'll have to put a breakpoint in the server methods to catch them there.

Example 5   Creating an ActiveX Control Using MFC

Objective

You would like to create and access an ActiveX control using MFC. An ActiveX Control is a COM object that supports a particular set of classes specified by the ActiveX standard.

Strategy

The term ActiveX started out as referring strictly to COM objects that created controls (custom list boxes, buttons, etc.) in your application — more specifically, controls written in MFC and C++ that could be used in Visual Basic applications. But as time progressed, ActiveX began to apply to non-controls as well — when you create a COM DLL or EXE server using Visual Basic you can only create it using the ActiveX standard. Both the Dialog Editor and ClassWizard make adding an ActiveX control to your dialog box templates fairly easy.

Steps

Initialize the COM DLL

Unlike other Win32 APIs, you need to tell the COM API — specifically OLE32.DLL — that you're going to be using it today. There is one of two calls you can make to do this shown in the next two steps. You must make this call not only when your application first starts, preferably in the InitInstance() of an MFC application, but also at the start of any thread your application creates.

1. Your application can call:

```
::CoCreateInstance(
     NULL        //reserved
);
```

OR

2. Your application can call:

```
::CoInitializeEx(
     NULL,                        //always NULL
     COINIT_APARTMENTTHREADED  //or COINIT_MULTITHREADED
     );
```

You also need to add _WIN32_DCOM to your project settings under "Preprocessor definitions" in order to get the prototype definition for ::CoInitializeEx() included in your compile.

The second API function allows for better control of your COM objects in a multitasking environment. Please refer to Chapter 3 for more on how a COM object behaves in a multitasking environment.

Example 5   Creating an ActiveX Control Using MFC   **125**

Add the Control to the Dialog Template

1. In the Dialog Editor, right click on the dialog template and pick the "Insert ActiveX Control" menu item. Then, pick the ActiveX control from the list of registered controls.

Handle Control Events in your Dialog Class

1. To handle events generated by the control in this template's dialog class, start by right clicking on the control in the dialog editor.
2. Select the control's ID from the "Class or object to handle" list box.
3. Double-click the event you want to handle from the "New windows/message events" list box and edit the handler.

Access the Control's Functionality from your Dialog Class

1. To access the control's methods and properties, start by opening the ClassWizard and select this template's dialog class.
2. Click on the Member Variables tab and select the control in the list box and click on "Add Variable". The ClassWizard will now automatically wrap this control in a class and add it to your project — not unlike the last example.
3. Use this member variable class like any MFC control class.

Access a Non-Control ActiveX Object

Occasionally, you'll run into a ActiveX object that isn't a control. To create and access its object:

1. Click on "Project", "Add to Project" and then "Components and Controls…" in the Studio's menu to open the "Components and Controls Gallery". Then, select the registered ActiveX class as before. A class wrapper for the control will again be created and added to your project, only now you can embed it in any of your other application classes.

2. After embedding the class, you will also need to call its `Create()` method which has the effect of going out and creating the COM object:

```
virtual BOOL Create(LPCTSTR lpszClassName,
    LPCTSTR lpszWindowName, DWORD dwStyle,
    const RECT& rect,
    CWnd* pParentWnd, UINT nID,
    CCreateContext* pContext = NULL)
```

You'll notice this is the exact same syntax of creating a window. However, you should create an invisible window since it's only used in this case to communicate between the ActiveX object and your application using window messages.

3. Call the methods and properties of this class as usual.

4. To handle events from this object, you will need to add an event sink map to the window you specified as the parent window when creating the object. You can do this manually, however, you can also cheat by adding this object to a dummy dialog template and performing the same steps above to add an event handler to that dialog's class and then cut and paste the resulting map into your real parent window. An event sink looks like this:

```
BEGIN_EVENTSINK_MAP(CTesterDlg, CDialog)
    //{{AFX_EVENTSINK_MAP(CTesterDlg)
    ON_EVENT(CTesterDlg, IDC_SERVERCTRL1, 1 /* Event1 */,
    OnEvent1Serverctrl1, VTS_I4)
    ON_EVENT(CTesterDlg, IDC_SERVERCTRL1, -600 /* Click */,
    OnClickServerctrl1, VTS_NONE)
    //}}AFX_EVENTSINK_MAP
END_EVENTSINK_MAP()
```

Note: Make sure to substitute your class's name and its base class for the dummy dialog class's name and base class.

Notes

- An ActiveX control is typically in a DLL, however, for marketing reasons, the suffix to this file is `.OCX` rather than `.DLL`.

Example 5   Creating an ActiveX Control Using MFC   **127**

- If you look inside of the class created for you by the Wizards, you'll notice that ActiveX objects use late binding. Perhaps the biggest difference between ActiveX and late bound objects is the standard functionality required of an ActiveX object. And an ActiveX control always creates a window of some sort. Please see the following listing for an example wrapper class.

- An ActiveX control is a great way to create a custom control once and share it amongst your project team without worrying about supplying them with the current source files or even whether or not they're programming in the same language as you.

- Events are a way for an ActiveX control to call back its client whenever something happens. This is done using windows messages and sink maps. Outside of the ActiveX standard, an entirely different approach is used where the server calls the client directly, which you will see in Examples 13 and 14 (beginning on page 160).

CD Notes

- You will need to create the ActiveX project first so that it registers the ActiveX control. Otherwise the Dialog Editor will complain that it can't find the control that was inserted into it and display a blank area where the control should be. Note that even before the application is created, its ActiveX controls are in use because some of its functionality is used by the Dialog Editor to configure the look and feel of the control (whatever settings the ActiveX author provides for their control.)

II   5

Listings — Generated C++ Class .H File

```
#if
   !defined(AFX_SERVER_H__A039CB7C_AE33_11D3_8088_000000000000__INCLUDED_)
#define
   AFX_SERVER_H__A039CB7C_AE33_11D3_8088_000000000000__INCLUDED_

#if _MSC_VER > 1000
#pragma once
#endif // _MSC_VER > 1000
// Machine generated IDispatch wrapper class(es) created by
   Microsoft Visual C++

// NOTE: Do not modify the contents of this file.  If this class is
   regenerated by
//  Microsoft Visual C++, your modifications will be overwritten.

/////////////////////////////////////////////////////////////////
   /////////
// CServer wrapper class

class CServer : public CWnd
{
protected:
   DECLARE_DYNCREATE(CServer)
public:
   CLSID const& GetClsid()
   {
       static CLSID const clsid
            = { 0xa039cb67, 0xae33, 0x11d3, { 0x80, 0x88, 0x0,
   0x0, 0x0, 0x0, 0x0, 0x0 } };
       return clsid;
   }
   virtual BOOL Create(LPCTSTR lpszClassName,
       LPCTSTR lpszWindowName, DWORD dwStyle,
       const RECT& rect,
       CWnd* pParentWnd, UINT nID,
```

Example 5   Creating an ActiveX Control Using MFC   **129**

```
             CCreateContext* pContext = NULL)
   { return CreateControl(GetClsid(), lpszWindowName, dwStyle,
   rect, pParentWnd, nID); }

   BOOL Create(LPCTSTR lpszWindowName, DWORD dwStyle,
      const RECT& rect, CWnd* pParentWnd, UINT nID,
      CFile* pPersist = NULL, BOOL bStorage = FALSE,
      BSTR bstrLicKey = NULL)
   { return CreateControl(GetClsid(), lpszWindowName, dwStyle,
   rect, pParentWnd, nID,
      pPersist, bStorage, bstrLicKey); }

// Attributes
public:
   long GetProperty1();
   void SetProperty1(long);
   short GetAppearance();
   void SetAppearance(short);
   float GetProperty2();
   void SetProperty2(float);

// Operations
public:
   long Method1(long lArg);
   void Refresh();
   void AboutBox();
};


//{{AFX_INSERT_LOCATION}}
// Microsoft Visual C++ will insert additional declarations
   immediately before the previous line.

#endif //
!defined(AFX_SERVER_H__A039CB7C_AE33_11D3_8088_000000000000__INCLUDED_)
```

II   5

Listings — Generated C++ Class .CPP File

```cpp
// Machine generated IDispatch wrapper class(es) created by
// Microsoft Visual C++

// NOTE: Do not modify the contents of this file.  If this class is
// regenerated by Microsoft Visual C++, your modifications will be
// overwritten.


#include "stdafx.h"
#include "server.h"


/////////////////////////////////////////////////////////////////////
    /////////
// CServer

IMPLEMENT_DYNCREATE(CServer, CWnd)

/////////////////////////////////////////////////////////////////////
    /////////
// CServer properties

long CServer::GetProperty1()
{
   long result;
   GetProperty(0x2, VT_I4, (void*)&result);
   return result;
}

void CServer::SetProperty1(long propVal)
{
    SetProperty(0x2, VT_I4, propVal);
}

short CServer::GetAppearance()
{
```

Example 5   Creating an ActiveX Control Using MFC   **131**

```
    short result;
    GetProperty(DISPID_APPEARANCE, VT_I2, (void*)&result);
    return result;
}


void CServer::SetAppearance(short propVal)
{
    SetProperty(DISPID_APPEARANCE, VT_I2, propVal);
}


float CServer::GetProperty2()
{
    float result;
    GetProperty(0x1, VT_R4, (void*)&result);
    return result;
}


void CServer::SetProperty2(float propVal)
{
    SetProperty(0x1, VT_R4, propVal);
}



/////////////////////////////////////////////////////////////////////
    ////////
// CServer operations

long CServer::Method1(long lArg)
{
    long result;
    static BYTE parms[] =
        VTS_I4;
    InvokeHelper(0x3, DISPATCH_METHOD, VT_I4, (void*)&result, parms,
        lArg);
    return result;
}


void CServer::Refresh()
```

II   5

```
{
    InvokeHelper(DISPID_REFRESH, DISPATCH_METHOD, VT_EMPTY, NULL, NULL);
}


void CServer::AboutBox()
{
    InvokeHelper(0xfffffdd8, DISPATCH_METHOD, VT_EMPTY, NULL, NULL);
}
```

Example 6   Creating an ActiveX Control from Visual Basic

Objective

You would like to create and access an ActiveX control from Visual Basic. An ActiveX Control is a COM object that supports a particular set of classes specified by the ActiveX standard.

Strategy

This is the classic example — one of the original forces behind the creation of COM, which is allowing Visual Basic programmers to access nifty controls created using MFC and C++. And as you might expect, as the oldest application of COM, it's very easy to accomplish using the Visual Basic studio.

Steps

Add an ActiveX Control to your VB Project

1. From the Visual Basic Studio, click on the "Project" then "Components..." menu commands to open the Components dialog box. Select the registered ActiveX control you require and click "OK". This control will then appear in the Studio's selection of controls.

2. Add the control to your form(s) and change its properties and handle its events as you would any regular VB control.

Example 7   Creating a COM Object from Visual Basic   **133**

Notes

- If the control you desire doesn't appear in the Components dialog box, chances are it hasn't been registered on your system. To register a control, you will first obviously need to locate the ActiveX file. Then, type:

```
Regsvr32 activex.ocx
```

where `activex.ocx` is the ActiveX file.

CD Notes

- Open the MFC application (under the Server subdirectory) using the Developer's Studio. Create the ActiveX control using the Developer Studio — the ActiveX control will automatically register itself as the last step of the project build. Open the VB application using the VB Studio and notice that this ActiveX control is now part of Form1.

Example 7   Creating a COM Object from Visual Basic

Objective

You would like to create and access a plain, ordinary COM object from Visual Basic.

Strategy

Visual Basic can also create COM objects that don't use the ActiveX standard and don't appear in its forms, however they still must support late binding. To do this, we will be using several additions to the Basic language lexicon, including `New`, `CreateObject()`, and even the `class.method()` syntax you only used to find in object-oriented languages.

Steps

Add the COM Class to Your Project

1. Add the COM object to your Visual Basic project by clicking on the "Project" and then "References..." menu commands to open the

"References" dialog box. From there, select the registered COM object you desire.

Create the COM Object

There are three ways to a COM object using VB, all with their own advantages and disadvantages:

1. Late binding is less efficient than early binding where method addresses are assigned to the client at compile time. However late binding can be optimized somewhat by using the following syntax, which gives VB a chance to look up a method's ID at compile time instead of at runtime:

```
Dim IWzdSrv1 As New SERVERLib.Wzd     'create object
IWzdSrv1.Method1 (1234)               'call its methods
```

Notice that you call the methods of the COM object as if it were a class method.

2. When an object type isn't known at compile time, such as when a user action determines what type of object to use, you need to use the following syntax. At runtime, VB is forced to query the method IDs of a COM object as required (hey, get a faster computer):

```
Dim IWzdSrv2 As Object
Set IWzdSrv2 = New SERVERLib.Wzd
IWzdSrv2.Method1 (4321)
```

3. You can even create a COM object without having first included it in your project by specifying its ProgID directly as seen here:

```
Dim IWzdSrv3 As Object
Set IWzdSrv3 = CreateObject("Server.Wzd")
IWzdSrv3.Method1 (9876)
```

4. In all cases, you don't have to worry about releasing a COM object in VB — it's done for you automatically.

Notes

- As you can see, COM objects are accessed just like the methods of any class in an object-oriented language. Later, in Example 27 (page 234), we will see that you can also create your own COM object using VB and the class nomenclature. So, in addition to the other benefits of COM, it also

Example 8   Creating a COM Object from Visual J++   **135**

brought a limited version of object-oriented programming to VB programmers.

- When working with a VB client and a C++ COM class, you'll find that you can't build the COM class if the VB project is open. Maybe in the next version…

CD Notes

- Create both the VB project and the ATL project (for debug). Unlike the situation where both the client and the COM object are both written in the same language, you cannot step from one language into another when debugging. You therefore have two choices, the first of which is to step through the VB client and watch as the objects are created and used. To test from the other side using C++, create the object as usual for debug. Also, create the VB client as an EXE file. Then, go into your C++ project settings and locate the "Debug" tab. On that page is an edit box called, "Executable for Debug Session". Click the button next to it and then click "Browse". Now locate the VB .EXE file you just created and select it. You can now start debugging in the Developer Studio which will automatically bring up the VB .EXE application.

Example 8   Creating a COM Object from Visual J++

Objective

You would like to create and access a COM object from Visual J++.

Strategy

Accessing a COM object from J++ is almost as simple as accessing it from Visual Basic — which is saying something. The Visual J++ Studio creates a Java class wrapper that takes all the headache out of creating, accessing, and destroying the object. Unfortunately, as with VB, you are again forced to use COM classes that support late binding.

Steps

Create the Class Wrapper

1. From within the VJ++ Developer Studio, click on the "Project" and "Add COM Wrapper..." menu items to open the COM Wrappers dialog box. You'll notice that the Studio creates two wrapper classes right in the COM class's subdirectory. The first class wraps the COM class's interface, the second wraps the COM class's implementation. Please see the following listings for an example of this wrapper class.

2. In the parlance of Java, the name of the DLL or EXE file that contains the COM class becomes the package name, but the class name is still the class name. Therefore, to create a COM class where the COM class "Wzd" exists in the DLL project "server", you would use:

```
server.Wzd wzd=new server.Wzd();
```

3. Call its methods as usual using:

```
wzd.Method1(1234);
```

4. As with Visual Basic, you don't have to worry about releasing a COM object — it's done for you automatically by VJ++.

Notes

The AppWizard will automatically add these menu items along with some context-sensitive help, so it's better to just manually add these menu items yourself than to add all that dead weight to your project.

CD Notes

Bring up both projects, the first in the VJ++ Studio, the second (located in the Server subdirectory) in the Developer Studio. Build both. Then, refer to the CD Notes for the last example.

Example 8   Creating a COM Object from Visual J++   **137**

Listings — Interface Wrapper Class

```
//
// Auto-generated using JActiveX.EXE 5.00.2918
//   ("C:\Program Files\Microsoft Visual Studio\VJ98\jactivex.exe"
// /wfc  /w /xi /X:rkc /l "C:\TEMP\jvcBA.tmp" /nologo /d
// "G:\jeswanke\mfc3\Examples\Creating a COM Object using Visual JPP"
// "G:\jeswanke\mfc3\Examples\Creating a COM Object using Visual
// JPP\Server\Server.tlb")
//
//  WARNING: Do not remove the comments that include "@com" directives.
//  This source file must be compiled by a @com-aware compiler.
//  If you are using the Microsoft Visual J++ compiler, you must use
//  version 1.02.3920 or later. Previous versions will not issue an
//  error but will not generate COM-enabled class files.
//


package server;

import com.ms.com.*;
import com.ms.com.IUnknown;
import com.ms.com.Variant;

// Dual interface IWzd
/** @com.interface(iid=C177116E-9AAA-11D3-805D-000000000000,
  thread=AUTO, type=DUAL) */
public interface IWzd extends IUnknown
{
  /** @com.method(vtoffset=4, dispid=1, type=METHOD,
    name="Method1", addFlagsVtable=4)
      @com.parameters([in,type=I4] lArg) */
  public void Method1(int lArg);
```

II   5

```
  public static final com.ms.com._Guid iid = new
    com.ms.com._Guid((int)0xc177116e, (short)0x9aaa, (short)0x11d3,
    (byte)0x5d, (byte)0x0, (byte)0x0, (byte)0x0, (byte)0x0,
    (byte)0x80,  (byte)0x0, (byte)0x0);
}
```

Listings — Implementation Wrapper Class

```
//
// Auto-generated using JActiveX.EXE 5.00.2918
//   ("C:\Program Files\Microsoft Visual Studio\VJ98\jactivex.exe"
    /wfc  /w /xi /X:rkc /l "C:\TEMP\jvcBA.tmp" /nologo /d
    "G:\jeswanke\mfc3\Examples\Creating a COM Object using Visual
    JPP" "G:\jeswanke\mfc3\Examples\Creating a COM Object using
    Visual JPP\Server\Server.tlb")
//
// WARNING: Do not remove the comments that include "@com" directives.
// This source file must be compiled by a @com-aware compiler.
// If you are using the Microsoft Visual J++ compiler, you must use
// version 1.02.3920 or later. Previous versions will not issue an
// error but will not generate COM-enabled class files.
//

package server;

import com.ms.com.*;
import com.ms.com.IUnknown;
import com.ms.com.Variant;

/** @com.class(classid=C177116F-9AAA-11D3-805D-000000000000,DynamicCasts)
    @com.interface(iid=C177116E-9AAA-11D3-805D-000000000000,
    thread=AUTO, type=DUAL) */
public class Wzd implements IUnknown,com.ms.com.NoAutoScripting,server.IWzd
{
  /** @com.method(vtoffset=4, dispid=1, type=METHOD,
    name="Method1", addFlagsVtable=4)
    @com.parameters([in,type=I4] lArg) */
  public native void Method1(int lArg);
```

Example 8   Creating a COM Object from Visual J++   **139**

```
public static final com.ms.com._Guid iid = new
  com.ms.com._Guid((int)0xc177116e, (short)0x9aaa, (short)0x11d3,
  (byte)0x5d, (byte)0x0, (byte)0x0, (byte)0x0, (byte)0x0,
  (byte)0x80,  (byte)0x0, (byte)0x0);

public static final com.ms.com._Guid clsid = new
  com.ms.com._Guid((int)0xc177116f, (short)0x9aaa, (short)0x11d3,
  (byte)0x80, (byte)0x5d, (byte)0x0, (byte)0x0, (byte)0x0,
  (byte)0x0, (byte)0x0, (byte)0x0);
}
```

6

Chapter 6

Writing COM Servers
with MFC

You will find in this chapter that writing a late-bound (aka, Automation) COM object using MFC is almost as easy as writing a similar object with Visual Basic or Visual J++. Unfortunately, the interface is just as slow, but writing an MFC COM class that supports early binding (aka, custom interface) can be a chore. Hopefully, however, the examples below will make this chore easier.

Why doesn't MFC make writing a custom interface easier? Why did all of that effort go into the ATL wizards instead, which does make creating a custom interface easier? True, MFC is more of a user interface library that doesn't quite count on the speed that ATL was designed for. And ATL doesn't have all the girth that a MFC application carries around, but then again, it doesn't have any of the niceties either. Whatever the reason, if your COM object does a lot of work with the user interface, you should use MFC and not ATL (you're back to talking to the Win32 API directly, for goodness sake.)

The examples in this chapter include:

Example 9   Writing an Interface Server Project   which is required for all MFC COM servers that want to support early-binding.

Example 10   Writing a COM DLL Server with MFC   which shows you how to create a COM class that lives in a DLL.

Example 11   Writing a COM EXE Server with MFC   which shows you how to write a COM class that lives in an executable.

Example 12   Writing a COM Server that Supports Late Binding with MFC   which shows you how easy it is to create a COM class that supports late binding.

Example 13   Writing a COM Server with a Connection Point with MFC   which shows you how to add support to your COM class so that it can call its client when something happens so that the client doesn't have to wait.

Example 14   Writing a COM Client with a Sink Using MFC   which shows the client side of the previous example.

Example 15   Writing a COM Singleton Server with MFC   which shows you how to write a COM class that only creates one instance of itself no matter how many times ::CoCreateInstance() is called — which is very useful when maintaining any constant data between other COM objects.

Example 16   Aggregating a COM Object with MFC   which shows you how to write a COM object that sort of derives functionality from another COM object.

Example 17   Writing an ActiveX Control Using MFC   which   shows you how to create a simple ActiveX control using MFC.

Example 9   Writing an Interface Server Project   **143**

Example 9   Writing an Interface Server Project

Objective

You would like to create an interface project that can be used to create an MFC COM class that supports early binding.

Strategy

MFC COM classes that will only support late binding (aka, Automation), which is all that Visual Basic and VJ++ can use, don't require an Interface Server. However, to write an MFC COM class that supports early binding (aka, a custom interface), you need an Interface Server. Why does the interface between a client and a server itself need a server? Because the job of sending and receiving data between client and server has been moved to this interface server. We will find in the next chapter that this separate project in MFC has been incorporated into the server project itself in ATL.

Because there aren't any wizards to create this project and it involves several intricate files, we will be simply modifying the Interface Server project found on the CD accompanying this book.

Steps

Add the Interface Server Project to Your MFC Server

1. Create a subdirectory off of your MFC COM server project and copy the following Interface Server project files from the CD. If you use the SampleWizard to do this, the "Wzd" mnemonic will be replaced with whatever is more appropriate for your application:

```
IWZD.DEF  // a DLL definition file
IWZD.DSP  // a project file
IWZD.DSW  // a workspace file
IWZD.IDL  // a blank IDL file
IWZD.MAK  // a make file that creates the .c, .h and .dll files
```

Note: You can find a sample listing for most of these files at the end of this example.

---

2. Insert this project into your COM server project's workspace.

Edit the IDL File

You will now need to add your own methods and method parameters to the IDL to define exactly what your COM class will be implementing. A first glance at an IDL can be rather imposing. However, when you look for a specific pattern, the syntax becomes clear to a C++ or Java programmer. In particular, the syntax is identical to that for defining a class except that every item you might find in a class definition can also have an attribute that further defines it. And these attributes are separated from the usual class definitions with braces ('[]'). What this means is that a class definition that might look like this:

```
Class MyClass : public CBase
{
    int Method1(long arg);
}
```

now looks like this:

```
[
class attributes
]
class MyClass :CBase
{
[method attributes] int Method1([argument attributes]long arg);
}
```

Next, we would substitute "interface" for "class" which defines an abstract class with all public methods. Also different is that the base class is always IUnknown for a COM object that doesn't support late binding. And finally, the return type for a COM object is off-limits to you and must be defined as HRESULT. COM uses this return value to return COM specific errors. The final product might look something like this:

```
[
    object,
    uuid(C177116E-9AAA-11D3-805D-000000000000),
    pointer_default(unique)
]
```

Example 9   Writing an Interface Server Project   **145**

```
interface IMyClass : IUnknown
{
    HRESULT Method1([in] long lArg);
};
```

Notice that there are no method attributes here — they usually come into play when defining a late-bound COM class.

To edit this file then:

1. Change the guid that defines this class. Use the `GUIDGEN.EXE` application that comes with your VC++ distribution kit to generate a new guid.

2. Change the class (aka, `interface`) name to your own class name.

3. Add your own methods and arguments to this class. To determine what attributes to add to these arguments, please refer to Example 31 (page 246).

4. If you need to define more than one COM class, simply cut and paste this definition as many times as needed and repeat steps 1 to 3.

Register the Interface

The MFC COM server that eventually uses this project needs to be registered with the system in order for `::CoCreateInstance()` to work. But if that MFC COM server is out-of-process, or if it's multitasking, you will also have to register the DLL generated by this project. When an MFC COM server is out-of-process, not only does COM use the Interface ID to request a particular COM class in that server, but it also uses it to load up this DLL to handle the communication between client and server. This communication also needs handling in multitasking servers when COM uses that communication for thread safety. Please see Chapter 3 for more on this topic.

1. The `.mak` supplied on the CD already automatically registers the DLL. To manually register this DLL use:

```
regsvr32 iwzdps.dll
```

or whatever the name of the DLL is.

Notes

- We will find in Example 18 (page 192) that this DLL registration isn't necessary with some COM objects written using ATL — even when they're out-of-process or using multitasking. That's because the COM

DLL is configured to do all the work itself as it does with objects written in VB and VJ++.

CD Notes

- Please refer to the next example for a demo of this project.

Listings — DLL Definition File

```
LIBRARY      "IWzd"

DESCRIPTION  'Proxy/Stub DLL'

EXPORTS
    DllGetClassObject    @1    PRIVATE
    DllCanUnloadNow      @2    PRIVATE
    GetProxyDllInfo      @3    PRIVATE
    DllRegisterServer        @4    PRIVATE
    DllUnregisterServer      @5    PRIVATE
```

Listings — Interface Defintion Language File

```
// IWzd.idl :  IDL source for the IWzd interface
//


// NOTE: no automation declarations—use MFC's built-in support for that
import "unknwn.idl";
    [
        object,
        uuid(C177116E-9AAA-11D3-805D-000000000000),
        pointer_default(unique)
    ]
    interface IWzd : IUnknown
    {
        HRESULT Method1([in] long lArg1, [out] long *plArg2);
        HRESULT Method2([in] long lArg, [in] unsigned long ulArg);
    };
```

Example 10   Writing a COM DLL Server with MFC   **147**

Listings — Make File

```
IFACE = IWzd

$(IFACE).dll: dlldata.obj $(IFACE)_p.obj $(IFACE)_i.obj
   link /dll /out:$(IFACE).dll /def:$(IFACE).def /entry:DllMain
   dlldata.obj $(IFACE)_p.obj \
     $(IFACE)_i.obj kernel32.lib rpcndr.lib rpcns4.lib rpcrt4.lib
   oleaut32.lib uuid.lib
   regsvr32 /s $(IFACE).dll

dlldata.c $(IFACE)_p.c $(IFACE)_i.c $(IFACE).h : $(IFACE).idl
   midl $(IFACE).idl

.c.obj:
   cl /c /Ox /DWIN32 /D_WIN32_WINNT=0x0400 /DREGISTER_PROXY_DLL $<

clean:
   @del $(IFACE).dll
   @del $(IFACE).lib
   @del $(IFACE).exp
   @del dlldata.obj
   @del $(IFACE)_p.obj
   @del $(IFACE)_i.obj
```

Example 10   Writing a COM DLL Server with MFC

Objective

You would like to write your own COM server using a dynamically linked library (DLL) that supports MFC and early binding (aka, custom interface).

Strategy

The AppWizard, ClassWizard, and MFC are all stream-lined to easily create a late-bound (automation) COM class for you. There isn't, however, an ounce of help available to add a custom interface to that. So we will use the wizards as far as they can take us and then manually edit the rest of the way.

We will be using the AppWizard to create the DLL server for us and the ClassWizard to create the COM classes themselves. We will also need the Interface Server found in Example 9 to define our COM classes for the world.

Steps

Create the Server Project

1. Use the AppWizard to create a "Regular DLL" using MFC statically or dynamically. Make sure to also include "Automation" support.

2. To a subdirectory of the create project, add the Interface Server project found in Example 9. Insert this project into your server project's workspace for convenience. Follow the steps in Example 9 to add your COM class's methods to the IDL file and build the project.

Create the COM Class

1. Use the ClassWizard to create a new class derived from `CCmdTarget` or any other MFC that's derived from `CCmdTarget`. Once you pick `CCmdTarget`, etc. as your base class, you will notice that the options at the bottom of the dialog box will become enabled. For the first class, choose "Createable by Type ID". A Class ID for your server will be generated and added to this class. For additional COM classes, you don't need another Class ID so you can simply pick "Automation".

Edit the Generated COM Class `.H` File

1. Include the `.h` file generated by the Interface Server project.

2. Below the `DECLARE_INTERFACE_MAP` macro, add an interface map that looks something like this:

```
BEGIN_INTERFACE_PART(WzdClass, IWzd)
    STDMETHOD_(HRESULT,Method1)(long, long *);
    STDMETHOD_(HRESULT,Method2)(long, unsigned long);
END_INTERFACE_PART(WzdClass)
```

where `IWzd` is the name you gave your COM class's interface and `Wzd-Class` is the name of the class that will actually implement the methods

Example 10   Writing a COM DLL Server with MFC   **149**

of your COM class. Within this map, declare each of the methods you will be implementing.

Edit the Generated COM Class .CPP File

1. To identify this class as the one that will implement the class defined in the IDL file, add the Interface ID to its .CPP file, preferably below the Dispatch map. You can find the Interface ID defintion in the *_i.c file generated by the Interface Server project and it looks something like this:

```
// add this line from idl project (iwzd_i.c)
const IID IID_IWzd =
   {0xC177116E,0x9AAA,0x11D3,{0x80,0x5D,0x00,0x00,0x00,0x00,0x00,0x00}};
```

2. Locate the implementation of the interface map and define your interface there, just below the definition of the IDispatch interface like so:

```
BEGIN_INTERFACE_MAP(CWzdSrv, CCmdTarget)
   INTERFACE_PART(CWzdSrv, IID_IWzdSrv, Dispatch)
   INTERFACE_PART(CWzdSrv, IID_IWzd, WzdClass) // add this line
END_INTERFACE_MAP()
```

Notice in this line that you are associating the Interface ID with the class name you specified in the .h file as the one that will be implementing the COM class.

3. You will now need to implement all of the methods of IUnknown for your COM class. Fortunately, there aren't many and other than their names, they'll look exactly like this:

```
///////////////////////////////////////////////////////////
/// XWzdClass Implementation
ULONG FAR EXPORT CWzdSrv::XWzdClass::AddRef()
{
   METHOD_PROLOGUE(CWzdSrv, WzdClass);
   return pThis->ExternalAddRef();//pThis accesses enclosing
   class's this pointer
}

ULONG FAR EXPORT CWzdSrv::XWzdClass::Release()
{
   METHOD_PROLOGUE(CWzdSrv, WzdClass);
```

II

6

```
    return pThis->ExternalRelease();
}


HRESULT FAR EXPORT CWzdSrv::XWzdClass::QueryInterface(REFIID iid,
    void FAR* FAR* ppvObj)
{

    METHOD_PROLOGUE(CWzdSrv, WzdClass);
    return (HRESULT)pThis->ExternalQueryInterface(&iid, ppvObj);
}
```

You'll notice that the name you pick for the COM class implementation (in this example `WzdClass`) gets an "X" prefixed to it by the macros and then is used to create a class that's embedded in your `CCmdTarget` class.

4. Similarly, add your COM class's methods to this class using the same naming convention and syntax as seen here:

```
STDMETHODIMP CWzdSrv::XWzdClass::Method1(long lArg1, long *plArg2)
{

    METHOD_PROLOGUE(CWzdSrv, WzdClass);


    return S_OK;
}
```

5. After building the project, you will need to manually register it using:

```
Regsvr32 xxx.dll
```

For a full listing of the `.cpp` and `.h` files, please refer to the listings at the end of this example.

Notes

- You can also modify your project settings to automatically register your COM object, however some prefer to always manually register their objects for more control over their development (you know exactly what COM object is registered).

- Don't mess with the other maps you find in the generated COM class files. Although they aren't used for a COM class that uses a custom interface, you might eventually decide to add late-binding support so that VB and VJ++ applications can use your creation. And these maps are where

Example 10   Writing a COM DLL Server with MFC   **151**

your definitions will go. The same applies to the .ODL file you'll find in your project directory.

- You create a Regular DLL project for the basic reason that a Regular DLL can be used by any application — not just an MFC application — which is one of the major tenants of COM: interoperability.

- Please see the notes under Example 18 (page 192) on how to debug your DLL server.

CD Notes

- Open the test.dsw project — all three projects will be loaded (the client, server and interface server). Build the debug version and place a break point in the OnTest() handler of the dialog class of the tester. Then, step through as the MFC client calls the MFC server.

Listings — COM Class .H File

```
#if
    !defined(AFX_WZDSRV_H__4487D433_A6FF_11D3_A398_00C04F570E2C__INCLUDED_)
#define
    AFX_WZDSRV_H__4487D433_A6FF_11D3_A398_00C04F570E2C__INCLUDED_

#if _MSC_VER > 1000
#pragma once
#endif // _MSC_VER > 1000
// WzdSrv.h : header file
//

#include "..\IServer\IWzd.h"

//////////////////////////////////////////////////////////////////////
    /////////
// CWzdSrv command target

class CWzdSrv : public CCmdTarget
{
    DECLARE_DYNCREATE(CWzdSrv)
```

II

6

```
    CWzdSrv();              // protected constructor used by dynamic
                            // creation

// Attributes
public:

// Operations
public:

// Overrides
    // ClassWizard generated virtual function overrides
    //{{AFX_VIRTUAL(CWzdSrv)
    public:
    virtual void OnFinalRelease();
    //}}AFX_VIRTUAL

// Implementation
protected:
    virtual ~CWzdSrv();

    // Generated message map functions
    //{{AFX_MSG(CWzdSrv)
        // NOTE - the ClassWizard will add and remove member
    functions here.
    //}}AFX_MSG

    DECLARE_MESSAGE_MAP()
    DECLARE_OLECREATE(CWzdSrv)

    // Generated OLE dispatch map functions
    //{{AFX_DISPATCH(CWzdSrv)
        // NOTE - the ClassWizard will add and remove member
    functions here.
    //}}AFX_DISPATCH
    DECLARE_DISPATCH_MAP()
    DECLARE_INTERFACE_MAP()
    BEGIN_INTERFACE_PART(WzdClass, IWzd)
```

Example 10   Writing a COM DLL Server with MFC   **153**

```
        STDMETHOD_(HRESULT,Method1)(long, long *);
        STDMETHOD_(HRESULT,Method2)(long, unsigned long);
    END_INTERFACE_PART(WzdClass)
};


/////////////////////////////////////////////////////////////////////
    ////////
```

Listings — COM Class .CPP File

```
//{{AFX_INSERT_LOCATION}}
// Microsoft Visual C++ will insert additional declarations
   immediately before the previous line.

#endif //
   0
// WzdSrv.cpp : implementation file
//

#include "stdafx.h"
#include "Server.h"
#include "WzdSrv.h"

#ifdef _DEBUG
#define new DEBUG_NEW
#undef THIS_FILE
static char THIS_FILE[] = __FILE__;
#endif


/////////////////////////////////////////////////////////////////////
////////
// CWzdSrv

IMPLEMENT_DYNCREATE(CWzdSrv, CCmdTarget)

CWzdSrv::CWzdSrv()
{
```

II

6

```
    EnableAutomation();

    // To keep the application running as long as an OLE automation
    // object is active, the constructor calls AfxOleLockApp.

    AfxOleLockApp();
}


CWzdSrv::~CWzdSrv()
{
    // To terminate the application when all objects created with
    // with OLE automation, the destructor calls AfxOleUnlockApp.

    AfxOleUnlockApp();
}



void CWzdSrv::OnFinalRelease()
{
    // When the last reference for an automation object is released
    // OnFinalRelease is called.  The base class will automatically
    // deletes the object.  Add additional cleanup required for your
    // object before calling the base class.

    CCmdTarget::OnFinalRelease();
}



BEGIN_MESSAGE_MAP(CWzdSrv, CCmdTarget)
    //{{AFX_MSG_MAP(CWzdSrv)
        // NOTE - the ClassWizard will add and remove mapping macros here.
    //}}AFX_MSG_MAP
END_MESSAGE_MAP()

BEGIN_DISPATCH_MAP(CWzdSrv, CCmdTarget)
    //{{AFX_DISPATCH_MAP(CWzdSrv)
        // NOTE - the ClassWizard will add and remove mapping macros here.
```

Example 10   Writing a COM DLL Server with MFC   **155**

```
    //}}AFX_DISPATCH_MAP
END_DISPATCH_MAP()


// Note: we add support for IID_IWzdSrv to support typesafe binding
//  from VBA.  This IID must match the GUID that is attached to the
//  dispinterface in the .ODL file.


// {4487D431-A6FF-11D3-A398-00C04F570E2C}
static const IID IID_IWzdSrv =
{ 0x4487d431, 0xa6ff, 0x11d3, { 0xa3, 0x98, 0x0, 0xc0, 0x4f, 0x57,
   0xe, 0x2c } };


// add this line from idl project (iwzd_i.c)
const IID IID_IWzd =
    {0xC177116E,0x9AAA,0x11D3,{0x80,0x5D,0x00,0x00,0x00,0x00,0x00,
    0x00}};


BEGIN_INTERFACE_MAP(CWzdSrv, CCmdTarget)
    INTERFACE_PART(CWzdSrv, IID_IWzdSrv, Dispatch)
    INTERFACE_PART(CWzdSrv, IID_IWzd, WzdClass) // add this line
END_INTERFACE_MAP()


// {4487D432-A6FF-11D3-A398-00C04F570E2C}
IMPLEMENT_OLECREATE(CWzdSrv, "Server.WzdSrv", 0x4487d432, 0xa6ff,
   0x11d3, 0xa3, 0x98, 0x0, 0xc0, 0x4f, 0x57, 0xe, 0x2c)


///////////////////////////////////////////////////////////////////
    ////////
// CWzdSrv message handlers


//////////////////////////////////////////////////////////////////
/// XWzdClass Implementation
ULONG FAR EXPORT CWzdSrv::XWzdClass::AddRef()
{
    METHOD_PROLOGUE(CWzdSrv, WzdClass);
    return pThis->ExternalAddRef();//pThis accesses enclosing
    class's this pointer
```

```
}

ULONG FAR EXPORT CWzdSrv::XWzdClass::Release()
{
    METHOD_PROLOGUE(CWzdSrv, WzdClass);
    return pThis->ExternalRelease();
}

HRESULT FAR EXPORT CWzdSrv::XWzdClass::QueryInterface(REFIID iid,
    void FAR* FAR* ppvObj)
{
    METHOD_PROLOGUE(CWzdSrv, WzdClass);
    return (HRESULT)pThis->ExternalQueryInterface(&iid, ppvObj);
}

STDMETHODIMP CWzdSrv::XWzdClass::Method1(long lArg1, long *plArg2)
{
    METHOD_PROLOGUE(CWzdSrv, WzdClass);

    return S_OK;
}

STDMETHODIMP CWzdSrv::XWzdClass::Method2(long lArg, unsigned long
    ulArg)
{
    METHOD_PROLOGUE(CWzdSrv, WzdClass);

    return S_OK;
}
```

Example 11   Writing a COM EXE Server with MFC

Objective

You would like to write your own COM class that lives in an MFC application (EXE).

Example 11   Writing a COM EXE Server with MFC   **157**

Strategy

As you might expect, the strategy here is not much different than it was for the last example. This time, however, rather than creating a DLL using the AppWizard, you create an EXE project. The ClassWizard creates the COM class as shown in the last example.

Steps

Create the COM EXE Server

1. Use the AppWizard to create an EXE project (MDI, SDI, Dialog — it doesn't matter). Make sure to specify "Automation" support.

2. To a subdirectory of the create project, add the Interface Server project found in Example 9. Insert this project into your server project's workspace for convenience. Follow the steps in Example 9 to add your COM class's methods to the IDL file and build the project.

Create the COM Class

1. Simply follow the steps in the last example to add a COM class to this server. The listings in that example show such a class.

2. To register this class, you only need to run the server EXE once. One of the first functions it calls will automatically register it in the system registry.

Notes

- If running a COM EXE is inconvenient as a method to registering your COM class, you can also write your own .reg file that will do the same. A template of such a .reg file might look something like this:

```
REGEDIT4

[HKEY_CLASSES_ROOT\CLSID\{000209FF-0000-0000-C000-000000000046}]
@="Microsoft Word Application"

[HKEY_CLASSES_ROOT\CLSID\{000209FF-0000-0000-C000-
    000000000046}\LocalServer32]
@="c:\\PROGRA~1\\MICROS~2\\OFFICE\\WINWORD.EXE"
```

CD Notes

- Debugging a COM EXE is a difficult proposition. Because the debugger within the Developer's Studio can only debug one process at a time, you will need to open a second Studio for the COM server. Create debug versions of both client and server and then start the server. You can now select break points in the server. Now, start the client and put a break point on the `OnTest()` button handler. Click "Test" and step into the server. You'll notice that you will automatically switch between debuggers as you go from client to server and back.

- You can also just start the client because when you step into the server, you automatically switch in a Studio that was executed just for the occasion. In fact, unless you start the COM server in the second debugger, a third debugger will be started — just loading the server up into a debugger isn't enough to make COM recognize it as the server you want to run.

- Make sure to kill each of these COM EXE servers as you go along — otherwise you won't be able to rebuild the server (you'll get a share protection error because the EXE is still using its program image).

Listings — COM Class

Please refer to last example.

Example 12   Writing a COM Server that Supports Late Binding with MFC

Objective

You would like to write your own MFC COM server that supports a late binding (aka, Automation) interface.

Strategy

As mentioned in Example 3 (page 116), the AppWizard, ClassWizard, and MFC are all setup to make the creation of a late bound COM class easy. Therefore, we will be using the AppWizard to create the server — either DLL or EXE — then, we will be using the ClassWizard to create the COM

Example 12 Writing a COM Server that Supports Late Binding with MFC **159**

classes. And instead of using the editor to add methods to our class, we will use the ClassWizard yet again.

Steps

Create the COM Server

1. Use the AppWizard to create a DLL or EXE project. The DLL must be a Regular DLL. The EXE can be MDI, SDI, or Dialog. If you choose SDI or dialog, every time you create a new COM object, a new instance of the EXE file is executed. With an MDI application, only one instance is executed. Also make sure to choose "Automation" support.

Create the COM Class

1. Use the ClassWizard to add a COM class to your project derived from `CCmdTarget` or any class that is itself derived from `CCmdTarget` (i.e., `CWnd`, `CDialog`). At this point, you'll notice that the controls at the bottom of this dialog become enabled. For the first class, you can pick "Createable by Type ID" because this adds a `ClassID` to your project. For other COM classes in the same server, you can pick "Automation".

Add Functionality to Your COM Class

1. To add methods to your COM class, open the ClassWizard again, select the new class name under "Class name" and then select the "Automation" tab. Then, click the "Add Method…" button to add your methods. Type the method name into "External name" and its arguments into "Parameter list" or "Return type". Notice that you'll only be able to specify parameter types that Automation supports through the COM DLL.

2. To add properties to your COM class, click the "Add Properties…" button. Properties are equivalent to a class's member variables. But because a client can't (and shouldn't) access the member variable of a server class, you call methods instead which by proxy reads and writes those member variables. The difference between selecting "Member variable" or "Get/Set method" is insignificant to the client. It must still access the variable with either a method prefixed with "Set" or "Get". However, for the server with "Get/Set" you get to override these methods directly.

II

6

With "Member variable" you are only notified when the variable in question changes (the "Notification function").

Notes

- The AppWizard will automatically add these menu items along with some context-sensitive help. It's better to just manually add these menu items yourself than to add all that dead weight to your project.

CD Notes

- Please see Chapter 1 and 2 for more on the difference in efficiency between early binding and late binding. As previously mentioned, you may not have a choice — if you want a VB or VJ++ programmer to be able to use your MFC COM object, you may be forced to use late binding.

Example 13   Writing a COM Server with a Connection Point with MFC

Objective

You would like to write a COM server that will allow a client to connect to it so that the server can call the client when some event occurs.

Strategy

Rather than tie up a client waiting for an event to occur at the server, it's much more efficient to give the server a function to call in the client when something happens. In Win32 parlance, this is called a *callback address*. Since COM is brand new and someone was bored, new terms were required. And since one of the past analogies for COM was that objects were like hardware components, hardware terms were used — thus, the new terms *connection point* and *sink*. What those terms really mean is that one or more clients can give a server object a callback address that the server will call when something happens. The address the server calls in a client is considered a sink, and a server that can do this is considered to have a connection point.

Unfortunately, a client can't just pass a callback address to a server for all the same reasons why we weren't calling the server directly ourselves.

Example 13   Writing a COM Server with a Connection Point with MFC   **161**

Instead, the client itself must implement a COM class and pass that address to the server. Fortunately, we'll leave that implementation to another example (Example 14, page 163).

For now, we just have to add a connection point to our MFC server. Although MFC has no wizards for automatically adding a connection point to a server, it does supply a couple of macros you can manually add to the server yourself.

Steps

Add a Connection Point to a COM Class

1. Create the COM server and class as usual with the AppWizard and ClassWizard as shown in the earlier examples in this chapter.

2. Add an `include` to the top of the new COM class's `.h` file:

```
#include "afxctl.h"
```

3. To the bottom of the `.h` file of this new class, add a connection map using the following macros and syntax:

```
DECLARE_CONNECTION_MAP()
BEGIN_CONNECTION_PART (CWzdSrv, CallBackCP)
    CONNECTION_IID(IID_IWzdSink)
END_CONNECTION_PART(CallBackCP)
```

The Interface ID (`IID_IWzdSink` in this example) will be of the interface that the client will be implementing — you'll be supplied this by the client "sink" interface and include it in this file either using the `guids.h` method or by including the definition directly. `CWzdSrv` in this example is the COM class you are adding this connection point to. The macros turn the name "CallBackCP" into a member variable called `m_xCallBackCP` defined using the MFC class: `CConnectionPoint`.

4. To the `.cpp` of your COM class, add the implementation of the connection map like so:

```
BEGIN_CONNECTION_MAP(CWzdSrv, CCmdTarget)
    CONNECTION_PART(CWzdSrv, IID_IWzdSink, CallBackCP)
END_CONNECTION_MAP()
```

II

6

5. To the .cpp and .h file, add a method that you can call from anywhere that will handle the actual call to the client. In this example, we're calling it "CallBackClients()". Since the CConnectionPoint class allows multiple clients to connect to this object for events, we will need to call every client in a loop like so:

```
void CWzdSrv::CallBackClients(long data)
{
    const CPtrArray *pConnections=m_xCallBackCP.GetConnections();
    int nConnections=pConnections->GetSize();
    for (int i=0;i<nConnections;i++)
    {
        IWzdSink *pWzdSink=(IWzdSink*)(pConnections->GetAt(i));
        pWzdSink->Callback(data);
    }
}
```

Notice that we use the m_xCallBackCP member variable to get an array of pointers to all clients and then call the COM object, IWzdSink, of each and every client. This COM class as well as its method, Callback() are both defined in the client's sink COM object.

Notes

- The connection point we just implemented only supports early binding. Since both Visual Basic and Visual J++ are both clients that require late binding, you can only use this solution with an MFC client. Please see the next example for an MFC client that uses this connection point. VB and VJ++ can connect to a connection point, however it must be implemented using IDispatch. The ATL Wizards allows you to automatically add a connection point that supports late binding — ironically, it's an early-binding connection point that you must manually add (see Example 25, page 220). If you can, use ATL when you need to support a late-binding connection point. Otherwise, please refer to the code automatically generated by Example 25 as a guide for adding this functionality to your MFC COM server.

- The routine CallBackClients() shown previously is intended to be used by any other class in the server when the client needs to be informed of some event. In addition, the parameter list can be just as extensive as any method call.

Example 14   Writing a COM Client with a Sink Using MFC   **163**

CD Notes

- Build the debug version of `test.dsw`. Then, stick a breakpoint in the `OnTest()` button handler and watch as the client creates the server, connects its sink to it, calls it and then gets called back.

Example 14   Writing a COM Client with a Sink Using MFC

Objective

You would like to connect to a server so that it can inform you of when some event occurs.

Strategy

This is the client side of the previous example using MFC. What we will be doing is setting up our own mini-server within this client, getting a COM pointer to it and sending it to the server to register and call whenever some event occurs. We will be creating our mini-server manually and then use MFC's `AfxConnectionAdvise()` static function to make the actual connection. Please refer to the previous example as to why this mini-server is called a sink in COM.

Steps

Create the Mini-Server's Interface Project

1. Referring back to Example 9 (page 143), create an interface that will define the COM class with which the server will use to call you back. This can be like any server interface with multiple arguments, etc.

2. Create a `guids.h` file containing the Interface ID of the Interface Project you created. In this example, `IWzdSink` is the interface's name:

```
#if !defined guids_h
#define guids_h

const IID IID_IWzdSink =
```

```
    {0x20050FE0,0xA719,0x11d3,{0xA3,0x98,0x00,0xC0,0x4F,0x57,0x0E,
    0x2C}};
```

```
#endif
```

Implement the Mini-Server in the Client

1. Pick a class in the client application that will implement this mini-server. Then, add an interface map to the class's .h file specifying the mini-server class and its methods. In this example, WzdSinkClass implements the IWzdSink interface, which contains one method called "Callback":

```
DECLARE_INTERFACE_MAP()
    BEGIN_INTERFACE_PART(WzdSinkClass, IWzdSink)
        STDMETHOD_(HRESULT,Callback)(long);
    END_INTERFACE_PART(WzdSinkClass)
```

2. In this class's .cpp file, implement WzdSinkClass including the "Classback" method:

```
ULONG FAR EXPORT CTesterDlg::XWzdSinkClass::AddRef()
{
    METHOD_PROLOGUE(CTesterDlg, WzdSinkClass);
    return pThis->ExternalAddRef();//pThis accesses enclosing
    class's this pointer
}


ULONG FAR EXPORT CTesterDlg::XWzdSinkClass::Release()
{
    METHOD_PROLOGUE(CTesterDlg, WzdSinkClass);
    return pThis->ExternalRelease();
}


HRESULT FAR EXPORT CTesterDlg::XWzdSinkClass::QueryInterface(REFIID
    iid, void FAR* FAR* ppvObj)
{
    METHOD_PROLOGUE(CTesterDlg, WzdSinkClass);
    return (HRESULT)pThis->ExternalQueryInterface(&iid, ppvObj);
}
```

Example 14   Writing a COM Client with a Sink Using MFC   **165**

```
HRESULT __stdcall CTesterDlg::XWzdSinkClass::Callback(long lArg)
{
    METHOD_PROLOGUE(CTesterDlg, WzdSinkClass);

    return S_OK;
}
```

Connect the Sink to the Server

1. Create the COM object that will be calling us back as usual:

   ```
   IWzdPtr pPtr;
   HRESULT hr=pPtr.CreateInstance(__uuidof(Wzd));
   ```

2. Then, get a pointer to our own mini-server's class:

```
IWzdSink *iWzdSink=NULL;
hr = m_xWzdSinkClass.QueryInterface (IID_IWzdSink, (void**)
   &iWzdSink);
m_xWzdSinkClass.Release(); // undo AddRef() performed by QI,
   unneeded when server is us
```

   Notice that we didn't actually have to create an instance of our mini-server because it was created when our client's class was created.

3. Now that we have a pointer to the server and a pointer to our mini-server, we can use MFC's `AfxConnectionAdvise()` to connect the two:

   ```
   DWORD dwConnectID;
   AfxConnectionAdvise(
               pPtr.GetInterfacePtr(),   // the real server
               IID_IWzdSink,             // our sink's IID
               iWzdSink,         // our sink's pointer
               FALSE,            // TRUE == increment our sink
                                 //    object's ref count
               &dwConnectID);
                         // identifier for AfxConnectionUnadvise()
   ```

The server can now call the client's `Callback()` method at will.

Unconnect from the Server

1. To unconnect your sink from the server, use:

```
AfxConnectionUnadvise(
            pPtr.GetInterfacePtr(),      // the real server
            IID_IWzdSink,                // our sink's IID
            iWzdSink,        // our sink's pointer
            FALSE,           // TRUE == decrement our sink
                             // object's ref count
            dwConnectID);
                    // identifier from AfxConnectionAdvise()
```

Notes

- Even though the client is implementing a small server within itself, it must still follow all the rules of COM (i.e., write and register a proxy/stub for non-automation argument types, etc.). After all, if the server is in Detroit, Detroit must still somehow get back to your client in Denver.
- ActiveX servers use an entirely different approach to calling back a client which involves a windows message sent to the client which has a sink map to process it. Please see Example 17 (page 184) for more.

CD Notes

- Build the debug version of test.dsw. Then, stick a breakpoint in the OnTest() button handler and watch as the client creates the server, connects its sink to it, calls it, and then gets called back.

Example 15   Writing a COM Singleton Server with MFC

Objective

You would like to write a COM class which only creates one instance of itself no matter how many times one or more clients create it.

Example 15   Writing a COM Singleton Server with MFC   **167**

Strategy

Classes that only create one of themselves are called *singletons*. But singletons break at least one rule of COM objects: when a client creates a COM object it will receive a fresh copy every time. Unfortunately, there are times when you want the same object referred to each time, such as when you want several COM objects to keep track of some state.

Writing a COM singleton server with MFC is a matter of overriding the class factory that MFC uses and perverting it for our own gains. A class factory is the actual part of a COM server that creates an instance of the COM object and MFC holds this functionality in a class called `COleObjectFactory`. In order to get at it, we also need to rewrite a couple of MFC macros because the name `COleObjectFactory` is actually burned into the standard macros.

Steps

Write a Singleton Class

1. Write your MFC COM class as usual. See the examples earlier in this chapter.

2. To the top of the `.h` file of this COM class, add your own definitions for the MFC OLE macros `DECLARE_OLECREATE` and `IMPLEMENT_OLECREATE`. We do this because the class name it uses for `COleObjectFactory` is hard-coded and we can't get to it otherwise to specify our own class name (which, in this example, is `CWzdOleObjectFactory`):

```
/////////////////////////////////////////////////////////////////////
    ////////
// Implement our own DECLARE_OLECREATE macros
class CWzdOleObjectFactory;

#define DECLARE_OLECREATE_WZD(class_name) \
public: \
    static AFX_DATA CWzdOleObjectFactory factory; \
    static AFX_DATA const GUID guid; \

#define IMPLEMENT_OLECREATE_WZD(class_name, external_name, 1, w1,
    w2, b1, b2, b3, b4, b5, b6, b7, b8) \
```

```
AFX_DATADEF CWzdOleObjectFactory
class_name::factory(class_name::guid, \
    RUNTIME_CLASS(class_name), FALSE, _T(external_name)); \
AFX_COMDAT const AFX_DATADEF GUID class_name::guid = \
    { l, w1, w2, { b1, b2, b3, b4, b5, b6, b7, b8 } }; \
```

3. Substitute these new macros for the current macros. In the .h file, find the DECLARE_OLECREATE macro and rename it to DECLARE_OLECREATE_WZD. In the .cpp file, it's the IMPLEMENT_OLECREATE macro for the IMPLEMENT_OLECREATE_WZD macro.

4. Implement this new derivation of COleObjectFactory at the bottom of this .h file:

```
///////////////////////////////////////////////////////////////////
    ////////
// Implement our own derivation of COleObjectFactory class factory
class CWzdOleObjectFactory : public COleObjectFactory
{
public:
    CWzdOleObjectFactory( REFCLSID clsid, CRuntimeClass* pRuntimeClass,
        BOOL bMultiInstance, LPCTSTR lpszProgID ) :
        COleObjectFactory(clsid,pRuntimeClass,bMultiInstance,lpszProgID)
            {};

    // singleton version of CreateInstance
    CCmdTarget *OnCreateObject()
    {
    // return static singleton object
        static CWzdSrv wzd;
        return &wzd;
    }
};
```

Notice that we've overridden the OnCreateObject() member function of COleObjectFactory and that we return a static instance of our COM class there. For a complete listing of this .h file, please refer to the end of this example.

Example 15   Writing a COM Singleton Server with MFC   **169**

Notes

- We couldn't define the new `CWzdOleObjectFactory` class at the top of the .h file for compiler reasons. `CWzdSrv` couldn't be used in our implementation until it was defined.

CD Notes

- Open the `test.dsw` project into the Developer Studio and build the debug version of the project. Place a breakpoint in the `OnTest()` button handler and run the application. Then, step through and notice that no matter how many times the COM class is created using `::CoCreateInstance()`, the same COM object instance is returned.

II

6

Listings — COM Class .H File

```
#if
    !defined(AFX_WZDSRV_H__4487D433_A6FF_11D3_A398_00C04F570E2C__INCLUDED_)
#define
    AFX_WZDSRV_H__4487D433_A6FF_11D3_A398_00C04F570E2C__INCLUDED_

#if _MSC_VER > 1000
#pragma once
#endif // _MSC_VER > 1000
// WzdSrv.h : header file
//
#include "afxctl.h"
#include "..\IServer\IWzd.h"


/////////////////////////////////////////////////////////////////////
    ////////
// Implement our own DECLARE_OLECREATE macros
class CWzdOleObjectFactory;

#define DECLARE_OLECREATE_WZD(class_name) \
public: \
    static AFX_DATA CWzdOleObjectFactory factory; \
    static AFX_DATA const GUID guid; \
```

```
#define IMPLEMENT_OLECREATE_WZD(class_name, external_name, l, w1,
    w2, b1, b2, b3, b4, b5, b6, b7, b8) \
    AFX_DATADEF CWzdOleObjectFactory
    class_name::factory(class_name::guid, \
        RUNTIME_CLASS(class_name), FALSE, _T(external_name)); \
    AFX_COMDAT const AFX_DATADEF GUID class_name::guid = \
    { l, w1, w2, { b1, b2, b3, b4, b5, b6, b7, b8 } }; \


/////////////////////////////////////////////////////////////////////
    ////////
// CWzdSrv command target

class CWzdSrv : public CCmdTarget
{
    DECLARE_DYNCREATE(CWzdSrv)

    CWzdSrv();          // protected constructor used by dynamic creation

// Attributes
public:

// Operations
public:

// Overrides
    // ClassWizard generated virtual function overrides
    //{{AFX_VIRTUAL(CWzdSrv)
    public:
    virtual void OnFinalRelease();
    //}}AFX_VIRTUAL

// Implementation
    virtual ~CWzdSrv();
protected:

    // Generated message map functions
    //{{AFX_MSG(CWzdSrv)
```

Example 15   Writing a COM Singleton Server with MFC   **171**

```
#define IMPLEMENT_OLECREATE_WZD(class_name, external_name, l, w1, \
    w2, b1, b2, b3, b4, b5, b6, b7, b8) \
    AFX_DATADEF CWzdOleObjectFactory \
    class_name::factory(class_name::guid, \
        RUNTIME_CLASS(class_name), FALSE, _T(external_name)); \
    AFX_COMDAT const AFX_DATADEF GUID class_name::guid = \
      { l, w1, w2, { b1, b2, b3, b4, b5, b6, b7, b8 } }; \


/////////////////////////////////////////////////////////////////////
    /////////
// CWzdSrv command target

class CWzdSrv : public CCmdTarget
{
    DECLARE_DYNCREATE(CWzdSrv)

    CWzdSrv();          // protected constructor used by dynamic creation

// Attributes
public:

// Operations
public:

// Overrides
    // ClassWizard generated virtual function overrides
    //{{AFX_VIRTUAL(CWzdSrv)
    public:
    virtual void OnFinalRelease();
    //}}AFX_VIRTUAL

// Implementation
    virtual ~CWzdSrv();
protected:

    // Generated message map functions
    //{{AFX_MSG(CWzdSrv)
```

II

6

```
        // NOTE - the ClassWizard will add and remove member
    functions here.
    //}}AFX_MSG

    DECLARE_MESSAGE_MAP()
    DECLARE_OLECREATE_WZD(CWzdSrv)

    // Generated OLE dispatch map functions
    //{{AFX_DISPATCH(CWzdSrv)
        // NOTE - the ClassWizard will add and remove member
    functions here.
    //}}AFX_DISPATCH
    DECLARE_DISPATCH_MAP()
    DECLARE_INTERFACE_MAP()
    BEGIN_INTERFACE_PART(WzdClass, IWzd)
        STDMETHOD_(HRESULT,Method1)(long, long *);
        STDMETHOD_(HRESULT,Method2)(long, unsigned long);
    END_INTERFACE_PART(WzdClass)
private:
    long m_lSave;


};
////////////////////////////////////////////////////////////////////
    ////////
// Implement our own derivation of COleObjectFactory class factory
class CWzdOleObjectFactory : public COleObjectFactory
{
public:
    CWzdOleObjectFactory( REFCLSID clsid, CRuntimeClass*
    pRuntimeClass,
        BOOL bMultiInstance, LPCTSTR lpszProgID ) :
        COleObjectFactory(clsid,pRuntimeClass,bMultiInstance,lpszProgID)

        {};

    // singleton version of CreateInstance
    CCmdTarget *OnCreateObject()
```

Example 16   Aggregating a COM Object with MFC   **173**

```
    {
    // return static singleton object
        static CWzdSrv wzd;
        return &wzd;
    }
};


////////////////////////////////////////////////////////////////////////
    /////////

//{{AFX_INSERT_LOCATION}}
// Microsoft Visual C++ will insert additional declarations
    immediately before the previous line.

#endif //
    !defined(AFX_WZDSRV_H__4487D433_A6FF_11D3_A398_00C04F570E2C__INCLUDED_)
```

Example 16   Aggregating a COM Object with MFC

Objective

When writing a new COM server in MFC, you would like to add function-ality from another existing COM object to your object no matter what lan-guage it's written in.

Strategy

You can derive one COM class interface from another (just use an existing interface for your interface's base class.) But you can't derive one fully implemented COM class from another. Why not? The biggest problem is that what is traditionally considered inheritance is achieved at compile-time when a big table of pointers is created to all of the methods in the class. COM classes can't interact at compile-time because they might be written in six different languages. Instead, they can only interact at runtime when any call to a base class method would require a runtime solution. Also consider that your base class might be in Cleveland while your derived class could be in Cincinnati. A runtime solution would possible go along these lines: You create a COM object `IWzd2` with a base class `IWzd`. When you go to call a method on `IWzd2`, COM must first check to see if the method you're calling is actually in `IWzd2` or if it's, in fact, a method in `IWzd` and then make the call there. Microsoft has been promising to eventually add this functionality, but it probably won't be any time soon based on how long they've promised it.

Another solution is encapsulation — where you just instantiate the base class from within your "derived" class and pass any method calls directly to the base class. However, the problem with this approach is that you must manually add every method you want from your base class to your derived class.

The most commonly used solution is called aggregation, where the derived class again instantiates the base class. The difference is that the client itself has to use `QueryInterface()` to get a pointer to the base class so that it can use its methods. In fact, all that aggregation gets you other than the automatic creation of two COM objects at once is that you are assured that both objects are in the same state. (Aggregation is used for transactions in COM+ for this reason.)

Again, there is no wizard to allow you to aggregate your MFC COM object to another COM object. But to be fair, ATL has no wizard either. Instead, aggregation is a sophisticated dance of manual editing, most of which involves creating and releasing one object from within another.

Steps

Write the COM Class That Will be Aggregated

1. Create the MFC COM class as usual using the ClassWizard, then add the following to the COM class's constructor:

```
// allow this object to be aggregated by another
EnableAggregation();
```

Example 16   Aggregating a COM Object with MFC   **175**

Write the COM Class That Will Aggregate The "Base" Class

1. Create a `guids.h` file containing the class and interface IDs of the object you wish to aggregate:

```
#if !defined guids_h
#define guids_h

const IID CLSID_IWzdAggSvr = { 0x603f1e0c, 0xa7d8, 0x11d3, 0xa3,
    0x98, 0x0, 0xc0, 0x4f, 0x57, 0xe, 0x2c };
const IID IID_IWzdAgg =
    {0x7ECEE460,0xA804,0x11d3,{0xA3,0x98,0x00,0xC0,0x4F,0x57,0x0E,0x2C}};


#endif
```

2. Create the second MFC COM class as usual using the ClassWizard.

3. To the `.h` file of this second class, add a member variable that will hold the pointer to the aggregated class's object:

```
private:
    IUnknown* m_punkWzdAggSvr;
```

4. Also add an override of one of `CCmdTarget`'s methods called `OnCreateAggregate()`:

```
// override for aggregation
virtual BOOL OnCreateAggregates();
```

5. To the `.cpp` file of the second class, include the `guids.h` file:

```
#include "guids.h"
```

6. Also initialize the new member variable to NULL:

```
// aggregated object(s)
m_punkWzdAggSvr=NULL;
```

7. Also implement the `OnCreateAggregates()` method by creating an instance of the aggregated COM object:

```
BOOL CWzdSrv::OnCreateAggregates()
{
    // create the aggregate object(s)
    HRESULT hr = ::CoCreateInstance(CLSID_IWzdAggSvr,
```

II

6

```
                                           GetControllingUnknown(),
                                           CLSCTX_INPROC_SERVER,
                                           IID_IUnknown,
                                           (LPVOID*)&m_punkWzdAggSvr);

    if (FAILED(hr))
        {
        m_punkWzdAggSvr = NULL;
        return FALSE;
        }

    return TRUE;
    }
```

8. In the `OnFinalRelease()` function of this second class, release the aggregated object:

```
        // release aggregate(s) we created
        m_punkWzdAggSvr->Release();
```

9. And finally, add an aggregate macro to the interface map:

```
INTERFACE_AGGREGATE(CWzdSrv, m_punkWzdAggSvr)
```

10. For a listing of this "derived" class, please refer to the end of this example.

Access the Aggregated Class Using the COM API

1. Create the "derived" second COM class as usual using `::CoCreateInstance()` as seen in Example 1.

2. To access the "base" class, you can't simply call it's methods using the same class pointer. Instead, you will need to use that pointer's `QueryInterface()` method to get a pointer to that class like so:

```
        IWzdAgg *iWzdAgg=NULL;
        hr=iWzd->QueryInterface(IID_IWzdAgg, (LPVOID*)&iWzdAgg);
        if (FAILED(hr))
            {
```

Example 16   Aggregating a COM Object with MFC   **177**

```
        _com_error err(hr);
        AfxMessageBox(err.ErrorMessage());
        return;
    }
```

where iWzd is the pointer to the "derived" class and iWzdAgg is a pointer to the "base" class.

3. You can then access the methods of these two classes as before:

```
    iWzdAgg->Method1(1234,&lArg);
    iWzd->Method2(lArg2, ulArg);
```

4. You must still release both objects — aggregation doesn't even take that burden away from you:

```
    iWzd->Release();
    iWzdAgg->Release(); // must still release aggregated interface
```

Notes

- Accessing the "base" class can be done much cleaner with smart pointers. Please refer to Example 2 (page 110) for an example of this.

- As mentioned previously, all aggregation gets you is the assurance that whatever data is in one object's member variables will match the state of the other object. For instance, aggregation is used in COM+ to keep track of transactions — the outer "derived" object keeps track of the transaction while the inner "base" class does the actual data source manipulation.

CD Notes

- Open the test.dsw project and build the debug version. Place a breakpoint in the OnTest() button handler and run the application. Then, step through as both the aggregated and aggregating COM objects are created and used.

II

6

Listings — Aggregating COM Class .H File

```
#if
   !defined(AFX_WZDSRV_H__4487D433_A6FF_11D3_A398_00C04F570E2C__INCLUDED_)
#define
   AFX_WZDSRV_H__4487D433_A6FF_11D3_A398_00C04F570E2C__INCLUDED_


#if _MSC_VER > 1000
#pragma once
#endif // _MSC_VER > 1000
// WzdSrv.h : header file
//


#include "..\IServer\IWzd.h"


/////////////////////////////////////////////////////////////////////
   ////////
// CWzdSrv command target

class CWzdSrv : public CCmdTarget
{
    DECLARE_DYNCREATE(CWzdSrv)

    CWzdSrv();          // protected constructor used by dynamic creation

// Attributes
public:

// Operations
public:

// Overrides
    // ClassWizard generated virtual function overrides
    //{{AFX_VIRTUAL(CWzdSrv)
    public:
    virtual void OnFinalRelease();
    //}}AFX_VIRTUAL
```

Example 16   Aggregating a COM Object with MFC   **179**

```
    // override for aggregation
    virtual BOOL OnCreateAggregates();


// Implementation
protected:
    virtual ~CWzdSrv();

    // Generated message map functions
    //{{AFX_MSG(CWzdSrv)
        // NOTE - the ClassWizard will add and remove member
    functions here.
    //}}AFX_MSG

    DECLARE_MESSAGE_MAP()
    DECLARE_OLECREATE(CWzdSrv)

    // Generated OLE dispatch map functions
    //{{AFX_DISPATCH(CWzdSrv)
        // NOTE - the ClassWizard will add and remove member
    functions here.
    //}}AFX_DISPATCH
    DECLARE_DISPATCH_MAP()
    DECLARE_INTERFACE_MAP()
    BEGIN_INTERFACE_PART(WzdClass, IWzd)
        STDMETHOD_(HRESULT,Method2)(long, unsigned long);
    END_INTERFACE_PART(WzdClass)
private:
    IUnknown* m_punkWzdAggSvr;
};


///////////////////////////////////////////////////////////////
    ////////

//{{AFX_INSERT_LOCATION}}
// Microsoft Visual C++ will insert additional declarations
```

II

6

```
    immediately before the previous line.

#endif //
    !defined(AFX_WZDSRV_H__4487D433_A6FF_11D3_A398_00C04F570E2C__INCLUDED_)
```

Listings — Aggregating COM Class .CPP File

```cpp
// WzdSrv.cpp : implementation file
//

#include "stdafx.h"
#include "Server.h"
#include "WzdSrv.h"

#include "guids.h"

#ifdef _DEBUG
#define new DEBUG_NEW
#undef THIS_FILE
static char THIS_FILE[] = __FILE__;
#endif

/////////////////////////////////////////////////////////////////
    ////////
// CWzdSrv

IMPLEMENT_DYNCREATE(CWzdSrv, CCmdTarget)

CWzdSrv::CWzdSrv()
{
    EnableAutomation();

    // aggregated object(s)
    m_punkWzdAggSvr=NULL;

    // To keep the application running as long as an OLE automation
    //    object is active, the constructor calls AfxOleLockApp.
```

Example 16   Aggregating a COM Object with MFC   **181**

```
        AfxOleLockApp();
}


CWzdSrv::~CWzdSrv()
{
    // To terminate the application when all objects created with
    // with OLE automation, the destructor calls AfxOleUnlockApp.

    AfxOleUnlockApp();
}


void CWzdSrv::OnFinalRelease()
{
    // release aggregate(s) we created
    m_punkWzdAggSvr->Release();

    // When the last reference for an automation object is released
    // OnFinalRelease is called.  The base class will automatically
    // deletes the object.  Add additional cleanup required for your
    // object before calling the base class.

    CCmdTarget::OnFinalRelease();
}



BEGIN_MESSAGE_MAP(CWzdSrv, CCmdTarget)
    //{{AFX_MSG_MAP(CWzdSrv)
        // NOTE - the ClassWizard will add and remove mapping macros
    here.
    //}}AFX_MSG_MAP
END_MESSAGE_MAP()


BEGIN_DISPATCH_MAP(CWzdSrv, CCmdTarget)
    //{{AFX_DISPATCH_MAP(CWzdSrv)
        // NOTE - the ClassWizard will add and remove mapping macros
```

II

6

```
    here.
    //}}AFX_DISPATCH_MAP
END_DISPATCH_MAP()

// Note: we add support for IID_IWzdSrv to support typesafe binding
//  from VBA.  This IID must match the GUID that is attached to the
//  dispinterface in the .ODL file.

// {4487D431-A6FF-11D3-A398-00C04F570E2C}
static const IID IID_IWzdSrv =
{ 0x4487d431, 0xa6ff, 0x11d3, { 0xa3, 0x98, 0x0, 0xc0, 0x4f, 0x57,
    0xe, 0x2c } };

// add this line from idl project (iwzd_i.c)
const IID IID_IWzd =
    {0xC177116E,0x9AAA,0x11D3,{0x80,0x5D,0x00,0x00,0x00,0x00,0x00,
    0x00}};

BEGIN_INTERFACE_MAP(CWzdSrv, CCmdTarget)
    INTERFACE_PART(CWzdSrv, IID_IWzdSrv, Dispatch)
    INTERFACE_PART(CWzdSrv, IID_IWzd, WzdClass)
    INTERFACE_AGGREGATE(CWzdSrv, m_punkWzdAggSvr)  //add this line
    so that MFC knows to look for additional methods in aggregated
    object
END_INTERFACE_MAP()

// {4487D432-A6FF-11D3-A398-00C04F570E2C}
IMPLEMENT_OLECREATE(CWzdSrv, "Server.WzdSrv", 0x4487d432, 0xa6ff,
    0x11d3, 0xa3, 0x98, 0x0, 0xc0, 0x4f, 0x57, 0xe, 0x2c)

////////////////////////////////////////////////////////////////////
    ////////
// CWzdSrv message handlers

////////////////////////////////////////////////////////////
/// Aggregate Implementation
BOOL CWzdSrv::OnCreateAggregates()
```

Example 16 Aggregating a COM Object with MFC **183**

```
{
    // create the aggregate object(s)
    HRESULT hr = ::CoCreateInstance(CLSID_IWzdAggSvr,
                                    GetControllingUnknown(),
                                    CLSCTX_INPROC_SERVER,
                                    IID_IUnknown,
                                    (LPVOID*)&m_punkWzdAggSvr);

    if (FAILED(hr))
    {
        m_punkWzdAggSvr = NULL;
        return FALSE;
    }

    return TRUE;
}


/////////////////////////////////////////////////////////////////
/// XWzdClass Implementation
ULONG FAR EXPORT CWzdSrv::XWzdClass::AddRef()
{
    METHOD_PROLOGUE(CWzdSrv, WzdClass);
    return pThis->ExternalAddRef();//pThis accesses enclosing
    class's this pointer
}

ULONG FAR EXPORT CWzdSrv::XWzdClass::Release()
{
    METHOD_PROLOGUE(CWzdSrv, WzdClass);
    return pThis->ExternalRelease();
}

HRESULT FAR EXPORT CWzdSrv::XWzdClass::QueryInterface(REFIID iid,
    void FAR* FAR* ppvObj)
{
    METHOD_PROLOGUE(CWzdSrv, WzdClass);
    return (HRESULT)pThis->ExternalQueryInterface(&iid, ppvObj);
```

II

6

```
}

STDMETHODIMP CWzdSrv::XWzdClass::Method2(long lArg, unsigned long
    ulArg)
{
    METHOD_PROLOGUE(CWzdSrv, WzdClass);

    return S_OK;
}
```

Example 17   Writing an ActiveX Control Using MFC

Objective

You would like to write your own ActiveX Control using MFC.

Strategy

Controls are child window-like buttons and list boxes. Using MFC, you can add your own functionality to the standard system supplied control windows, but the classes you create would only work in an MFC application. ActiveX controls are COM objects that support a standard set of functionality so that you can add your custom controls to any other application that supports ActiveX controls.

What this example hopes to show is how to create a basic ActiveX control using MFC's ActiveX ControlWizard. But because ActiveX represents a whole standard for a server, including dozens of classes, it must support in order to be considered ActiveX, if you will be doing serious ActiveX development, I would advise you to get a book that can devote all of its pages to ActiveX.

Example 17   Writing an ActiveX Control Using MFC   **185**

Steps

Create the ActiveX Project

1. Click on Developer Studio's File and New menu items to open the New dialog box. Then, enter a project name and directory and select "MFC ActiveX ControlWizard".

2. The ControlWizard has two steps. In the first step, pick how many controls this server will contain. If you have two or more related controls, pick that many. The final DLL will be that much bigger, but there won't be any wasted space. You can also have the wizard provide everything you need so that your customers must buy a license from you in order to run this control. Please see Figure 6.1 for step 1.

3. In the next step, you pick the name to give this control's classes. You can also decide whether your control will be based on a standard system control, such as a button or list box (recommended), or whether you're going to create your own control from scratch. Also, in step 2 you can decide on several features to add to your control. Those shown in this page are related to visual features you can add. Press the "Advanced..." button to pick from a selection of options that can optimize your control when used in a web browser. Please see Figures 6.2 and 6.3 respectively for step 2 and the Advanced options. Please refer to Table 6.1 for the effect of each of these options.

Figure 6.1   Step 1 of ControlWizard

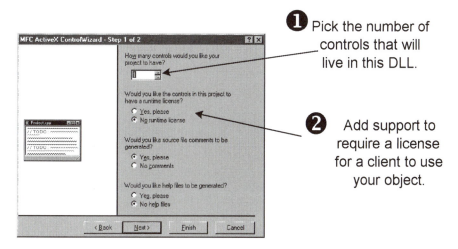

Figure 6.2   Step 2 of ControlWizard

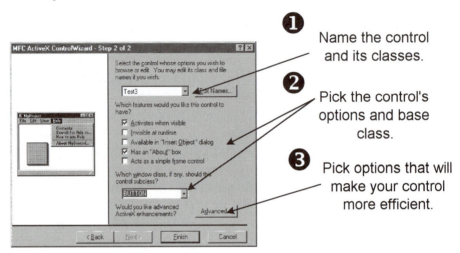

① Name the control and its classes.

② Pick the control's options and base class.

③ Pick options that will make your control more efficient.

Figure 6.3   Advanced ActiveX Features

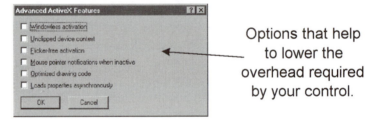

Options that help to lower the overhead required by your control.

Example 17   Writing an ActiveX Control Using MFC   **187**

Table 6.1   Advanced ActiveX Features

Activates when visible	All controls draw themselves and process messages. ActiveX controls have the additional ability to save system resources by going into an inactive state when the user isn't pounding on them. In this state, they can still draw themselves, but they don't process messages. If you never want your control to go into this dormant state and potentially present a delay whenever your user clicks on it, pick this option. If the client still wants to make your control inactive, it will.
Invisible at runtime	Pick this option for a control that really only does background processing, such as a timer control. Again, this is a request to the client — it may still choose to display your control.
Available in "Insert Object" dialog	Adds functionality to your control that allows the client to list the control in its "Insert Object" dialog box — only certain clients such as Microsoft Word implement this functionality.
Has an About box	Adds functionality to your control that allows it to display an about box whenever the client requests one.
Acts as a simple frame control	Adds the implementation for the `ISimpleFrameSite` ActiveX class so that your control can intercept messages from other ActiveX controls sitting in the client. Useful when implementing a group of controls such as a collection of radio buttons.
Windowless activation	Prevents your control from creating a system window (using the Win32 API call `::CreateWindow()`) to avoid all that overhead. Instead, you and your client are responsible for everything the system window used to do for you including drawing the control, determining if it has input focus and determining if the mouse cursor is hovering over it.
Unclipped device context	When drawing your control, the device context passed to you is clipped to allow you to only draw where your control should be. However, clipping the context adds overhead. So, if you promise not to draw where you're not supposed to, pick this option.

II

6

Flicker-free activation	If the look of your control won't change between when it's active (the user is pounding on it) or inactive, choose this option.
Mouse pointer notifications when inactive	Allows your control to still process certain mouse messages even when it's inactive (see the definition for active in the first row of this table).
Optimized drawing code	Allows you to be a pig with your device contexts. In other words, you don't have to worry about continually restoring the original graphic object to the device context — the client will restore everything for you when your control is done drawing, saving time on individual restores. However, not all clients support this option.
Loads properties asynchronously	Useful when your control is used in an Internet application, allows your control to be used right away by the client before all of its properties have been loaded from wherever. Of course, the client may find that it will have to wait anyways when the property it wants hasn't come through yet.

Write the ActiveX Control

1. Use the ClassWizard to add message handlers to the control class and implement your control as you would any custom windows control using MFC. For drawing, you can use the MFC CDC class. You might even consider referring to my previous books on MFC.

2. You can also add methods and properties to your ActiveX control. Because your control implements a late binding (aka, Automation) COM interface, you can use the "Automation" tab of the ClassWizard to add these methods as seen in Example 12. In addition to any custom methods or properties you might add however, you'll notice you can also add the stock methods and properties supported by the ActiveX standard.

3. You can now also click on the "ActiveX Events" of the ClassWizard to add events to your control. Events allow an ActiveX control to tell a client that something happened (i.e., the user clicked on it). Events are like the connection points and sinks we used in Examples 13 and 14 to allow a server to call a client, but unlike connection points, ActiveX controls send window messages to a client's event message map.

Example 17   Writing an ActiveX Control Using MFC   **189**

Write the ActiveX Control Properties

When you use the dialog editor to add controls to a dialog template, you can open a property sheet for each of those controls to change how a control will look and act. As an example, you have several choices when it comes to how a push button will look. By the same token, your ActiveX control can also support a property sheet that allows programmers using your control to change its initial look and feel in the following steps:

1. Use the ClassWizard's "Automation" tab to add any properties to your control class your "programmer user" can change, such as what color your control will be.

2. When your ActiveX project was created, it also created a property page class and dialog template. Use the Dialog Editor to open that template and add a control for every property you added in the last step. For color choice you might want to add a combo box listing every color allowable.

3. Back in the ClassWizard, select the property page class and click on the "Member Variables" tab. Then, add a member variable for every property control you added to the template. The name you give each of these properties must be the same as the name you gave it above in step 1.

4. Your programmer user (friend?) will now be able to access your property member variables from Visual Basic or even the Dialog Editor in the same way they could access the standard properties (see below). But for them to be able to save their settings, you will also need to add a macro for each property to the `DoPropExchange()` function back in your control's `.cpp` file. The actual macro name depends on the type of property you are saving. As an example, the following will save a long:

```
void CServerCtrl::DoPropExchange(CPropExchange* pPX)
{
    ExchangeVersion(pPX, MAKELONG(_wVerMinor, _wVerMajor));
    COleControl::DoPropExchange(pPX);

    PX_Long( pPX, "Property2", m_property2, 1234 );

}
```

Other properties include: `PX_Short`, `PX_Float`, `PX_Bool`. See your documentation for more.

Notes

- Please refer to the examples in Chapter 5 for how to use an ActiveX control from MFC or Visual Basic.

- While perfecting your control, you may find it too cumbersome to create and use another application to be your client. Instead, you can use the "ActiveX Control Test Container". To invoke this utility, just build the debug version of your control and then press the "Debug"/"Go" menu commands or toolbar button to start debugging. The debugger will immediately realize you need an executable to run your control (because it's a DLL) and will prompt you for one. Click on the arrow button next to the edit box and select the "ActiveX Control Test Container".

- In ActiveX parlance, a container is a client that also usually supplies the parent window on which your control will appear. Internet Explorer, Visual Basic forms, and even your dialog boxes can be containers.

CD Notes

- The project on the accompanying CD contains a simple ActiveX control. Try building it and inserting it in a Visual Basic form or VC++ dialog template.

Chapter 7

Writing COM Servers with ATL

As we saw in the last chapter (if you bothered with it), writing a COM server that supports early binding (aka, a Custom Interface) involves a lot of manual editing. But MFC was really designed as a way to wrap the Win32 API in order to create applications. The Active Template Library (ATL) on the other hand was designed with just one part of the API in mind — that which supports COM. By creating a library of classes that just support COM, you can create smaller and faster COM objects. The examples in this chapter almost mirror those for MFC in the last chapter, and if you compare them you'll find that the wizards and macros and classes in ATL almost make creating a COM object that supports both a custom interface *and* late binding as easy as creating an MFC class that only supports late binding.

The examples in this chapter include:

Example 18   Writing a COM DLL Server Using ATL   where we will look at how to create a COM DLL using ATL's wizards.

Example 19   Writing a COM EXE Server Using ATL where we do the same for COM EXE's.

Example 20   Writing a COM EXE Service Using ATL where we do the same for NT/Win2000 Services.

Example 21   Extending Your ATL COM Class where we examine how to add additional methods to an existing COM class while still supporting clients that used the old methods.

Example 22   Writing an ATL Server that Supports Late Binding where we discover how easy it is to support both early and late binding with ATL.

Example 23   Writing an ATL Singleton Server where we look at creating a COM class that only creates one object from itself to allow several COM objects to share data.

Example 24   Writing an ATL COM Tearoff Server where we discover how to create a Tearoff Interface, a way to create a smaller COM object by leaving infrequently used functionality out on the disk.

Example 25   Writing an ATL COM Server that has a Connection Point where we look at adding functionality to our COM class so that it can call back the client when some event occurs.

Example 26   Aggregating a COM Object Using ATL where we look at adding the weak version of class inheritance to our ATL COM class.

Example 18   Writing a COM DLL Server Using ATL

Objective

You would like to write your own COM server using a dynamically linked library and the Active Template Library (ATL).

Example 18   Writing a COM DLL Server Using ATL   **193**

Strategy

We will be using the ATL COM AppWizard to create the DLL project itself, but to create the COM classes that will live in this server, we will use the ATL Object Wizard. We will also look at adding methods and properties to a COM class using the `ClassView` tab of the Workspace View of the Studio.

Steps

Create the DLL Project

1. Use the Developer Studio's "File" then "New" command to open the New dialog box. Select the "Projects" tab and pick a name for your server. Remember this will be the name of the server and not the COM class that will live in it, and for the purposes of file naming you can't name both the same. Click on the "ATL COM Wizard" list box item to open the ATL COM Wizard. The DLL option is already selected, but you also have a choice of three optional types of functionality you can add to your server:

   - Allow merging of proxy/stub code — if your COM class's methods will be using non-standard argument types as seen in Example 31 (page 246), you will need to write and register a proxy/stub DLL to support those method arguments. This normally means you need to keep track of two DLLs, one for the COM server and one for the proxy/stub server. Clicking this option consolidates the COM DLL with the proxy/stub DLL so you have only one DLL to worry about.

   - Support MFC — if you would like to use MFC from your COM server (and who doesn't), click this option. Seriously though, don't click this option unless you will be doing some extensive user interface work in your server because it adds overhead to your object. If you need MFC's data collection support, why not use the Standard Template Library (STL) instead? That's what it's there for.

   - Support MTS/COM+ — if you would like to add support in your server for the ability to work in a COM+ environment allowing your object to work with several databases while COM worries about committing or rolling back the transaction then pick this option.

   Please refer to Figure 7.1 for a look at these options.

II

7

Figure 7.1   The ATL COM AppWizard

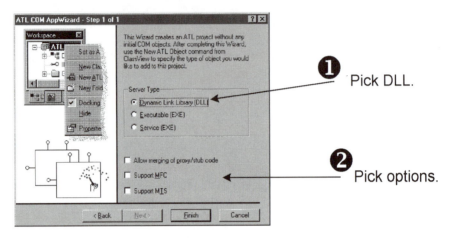

Add a COM Class to the Project

So far, all we've done is create a shell that our COM class will live in. Next we'll add a COM class to this shell.

1. Use the Studio's "Insert" then "New ATL Object" menu commands to open the ATL Object Wizard. Pick "Simple Object" and click "Next" to open the ATL Object Wizard Properties sheet.

Note: There are several other types of objects to choose here; however, 90% of the time you'll either picking this one or MS Transaction which is covered in Example 36 (page 276).

---

2. Fill in a name to give your COM class, which you will find will create several other names as well.

3. Click on the "Attributes" tab to view the options you can give your class.:

   • Threading model — allows you to decide whether or not to depend on COM to make sure your object is thread safe. Please refer to Chapter 3 for much more on what these options mean.

   • Interface — allows you to pick whether or not your object will support early or late binding. If you pick "Custom interface", Visual Basic and Visual J++ will not be able to use your object. Picking "Dual" allows other C++ objects and clients to use the

Example 18   Writing a COM DLL Server Using ATL   **195**

more efficient early binding to connect to your object *and* also allow VB and VJ++ to use your object as well — invisible to you. You do however have to make sure none of your methods use unauthorized argument types. If you get a warning about OLE Automation when this project's IDL file is being compiled with MIDL, you know the argument type you picked is wrong.

- Aggregation — allows you to decide if your object will allow another object to aggregate it or even if aggregation is required. Aggregation is COM's weak version of class derivation. For much more on it, please refer to Example 26 (page 226) or Chapter 3.

- Supports ISupportErrorInfo — tells a potential client that your object will be using SetErrorInfo() to save the any error information and that that client can call GetErrorInfo() to retrieve that information. Please see Example 46 (page 333) for how to use these two COM API calls.

- Support Connection Points — adds support to your object to allow it to talk back to a properly configured client. Please see Example 25 (page 220) for how to do this.

- Free Threaded Marshellar — really only applicable when using the "Both" threading model, this option prevents COM from ever adding thread protection to your object. If you pick a "Both" threading model, COM may or may not add thread protection to your object depending on what object uses it. But because you're already required to put your own thread protection in your object (using critical sections, etc.) in the event it is required, it doesn't make sense for COM to also add thread protection. So this option prevents COM from ever adding thread protection and removes that overhead. Please see Chapter 3 for much more on this topic.

Please refer to Figure 7.2 for a look at these options.

II

7

Figure 7.2   ATL Object Attributes

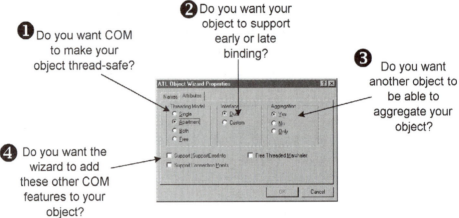

Add Methods and Properties to Your COM Class

Now we have a COM class with no methods or properties. You could manually add these items yourself — and believe me, you will eventually edit them anyways — but for a quick head start, you can automatically add the definitions for these items using the `ClassView` of your Workspace View.

1. Click on the `ClassView` tab of your Workspace View and locate the interface symbol for your COM class. Then, right-click on this symbol. You will then be given a choice of adding methods or properties to this class. Please refer to Figure 7.3.

Note: If you pick the symbol above the interface symbol in Figure 7.3, you will also be given an opportunity to add methods *but* you will be adding them to the DLL shell that your COM class sits in and not to the COM class itself.

---

2. To add a method, pick the "Method" menu item to open the "Add Method to Interface" dialog box. Enter the method's name and arguments. Note that you must enter how those arguments will appear in the IDL file, that is, with the argument attributes before the argument and type (i.e., `[in] long lArg`).

Example 18   Writing a COM DLL Server Using ATL   **197**

Figure 7.3   Adding Methods Using the ClassView

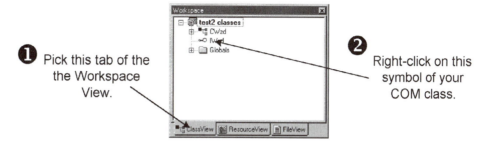

❶ Pick this tab of the the Workspace View.

❷ Right-click on this symbol of your COM class.

3. When you click "OK", the Studio will edit three files for you: the IDL file, and the .CPP and .H files of the COM class itself. Unfortunately, it will use the same argument types you entered in the IDL file, which may not work in the .CPP and .H files. In other words, you may be required to go back and fix the types so that your source files will compile. This most notably happens with Safearray arguments. Please refer to Chapter 2 for what arguments to use in the IDL and the C++ source files.

- You can now go to the .CPP file and implement you COM class's methods.

Build the Project

1. Although you have only one Debug build configuration, you have several Release configurations including minimum size or dependencies, with or without Unicode. Both size configurations simply determine whether you link to the ATL DLL file statically or dynamically — just like MFC. If you choose minimum size, then your server will try to link to the ATL DLL on your system at runtime. Minimum dependencies will cause your server to link to the ATL library instead at link time. If you want to avoid version problems with where your server will be installed, pick by dependencies. However, if you will have several server DLLs on one system, pick by size and also install whatever ATL DLL that you built against.

2. Unicode allows your server to support any character set, but you can't even build using this configuration unless your operating system supports Unicode (e.g., Windows NT or 2000 but not Windows 9x).

3. Another concern of Release building is that you will occasionally get an unresolved external symbol _main link error. You will only get this error if your server is using a C runtime library function (CRT) that requires startup code from your dll or exe, and your ATL project removed this

code to make your server smaller. To force ATL to return this code, go into your project settings and remove the symbol "_ATL_MIN_CRT" from your preprocessor definitions (you will only find this symbol in your Release configurations).

4. If your methods will be using argument types that aren't supported by the OLE dll (i.e., structures) and you want COM to either provide thread protection to your COM server or your COM server will be out-of-process as with an MTS/COM+ server, you will need to build your server's proxy/stub dll too. To build this dll, execute this line:

```
nmake -f XXXps.mk
```

Where XXXps.mk is the make file that the ATL COM AppWizard automatically created for your project. Because nmake requires thousands of utilities and dlls to be in your environments PATH, the easiest way to execute this line is to temporally add it to the Post Build Step of your project settings and rebuild the project. Once built, you need to register it, now and wherever your server is installed, using regsvr32.exe:

```
Regsvr32 XXXps.dll
```

If you don't create this DLL, COM is notorious for not telling you where it hurts and will try to muddle through without it. In the case of an array, for instance, it will send the first element over and ignore the rest.

5. An ATL project will automatically register itself when you build it. Some feel that this is too much automation on the project's part and go into the project settings (the "Custom Build" tab) and remove this step. However, I have found that 75% of my persistent COM problems when developing are from incorrectly registering the right COM server. Unless you feel you're an advanced user, I would leave this step alone.

Notes

- We were able to automatically add methods to our COM class using the ClassView tab of the Workspace View — however, if you make a mistake and want to add or change an argument, or if you want to delete a method or property entirely, you will have to manually edit the IDL, .CPP and .H files yourself. Also be aware that if the definition for a method doesn't match in all three files, you don't get a friendly compiler error telling you the problem. Instead you get an obscure error that the COM class you are writing couldn't be instantiated — at compile time. Before you waste your time pondering why the compiler was trying to

Example 19 Writing a COM EXE Server Using ATL **199**

instantiate your class, just recheck your prototypes against the IDL file to make sure they match both for argument types as well as number of arguments.

- A common scenario is to have an MFC client with one or more ATL server dependencies in the same workspace. But when you build the client using its sole "Release" build configuration, which build configuration does the Studio use for your ATL servers? Minimum size. If you prefer minimum dependencies instead, you can add a plain old "Release" configuration to your project which the Studio will use instead. Start by clicking on the "Build" then "Configuration" menu items to open the Configuration dialog box. Then, click "Add". Enter the name "Release" and copy it from the minimum dependencies configuration.

- It is a little trickier to debug a COM DLL than it is to debug a regular DLL. The problem is in setting breakpoints. Since a COM object doesn't officially become part of your application until runtime, the debugger will refuse to allow you to set a breakpoint in a COM DLL until runtime, and then when your application terminates, those breakpoints get thrown out. To make your COM DLL accept breakpoints as readily and dependably as a regular dll, go into your project settings under the "Debug" tab, and open the "Category" combo box at the top. Select "Additional DLLs" and browse for and select the .DLL files that contain the COM servers you want to debug. Now you can go into that server's source files and set breakpoints whether you're running or not.

Example 19 Writing a COM EXE Server Using ATL

Objective

You would like to write your own COM server running from an application written with the Active Template Library (ATL).

Strategy

As you might expect, writing the executable version of a COM server isn't that much different than writing the dll version which we did in the last example. However, there are a few things to keep in mind which we will discuss below.

Steps

Create the EXE Project

1. As in the last example, bring up the ATL COM AppWizard, only this time select "Executable (EXE)" for the server type. You will notice that all of the options listed at the bottom of the dialog box now become disabled. Why can't you merge the proxy/stub with your executable? Because the proxy/stub must be in-process with its client and if its sitting in your server's executable, it will be out-of-process. Why can't your executable server support MTS/COM+? Because those technologies depend on your server being run by their DLL host, which pretty much rules out an EXE. And why can't your server support MFC? Actually it can, and if you're willing to hunt around the MSDN documentation for it, they'll give you some tedious steps to manually add MFC support to your COM executable server. But I think it would be much simpler to refer to Example 11 (page 156) and create your COM executable server using MFC instead.

Create COM Classes and Add Methods

1. Please refer to the last example for how to create your COM classes and add methods. Pay particular notice to the step on building the proxy/stub. When building an executable server, you aren't forgiven as many sins as you are when using an in-process dll server. In particular, if one of your methods has a non-standard argument type and you don't build the proxy/stub dll, you're calling the method directly anyway so it doesn't matter. You will always be dependent on COM marshalling when you write a COM EXE server.

Build the Project

1. Please refer to the last example for the issues related to, among other things, what Release build configuration to use and how to alleviate the missing external symbol `_main` link error.

Register the Project

1. An ATL EXE project will automatically register itself when you build it. However, when the server is on a different machine than the client, you

Example 19   Writing a COM EXE Server Using ATL   **201**

will need to use the OLE/COM Object Viewer to configure the system registry on the client's machine to tell it what machine the server is on. Start by loading up the OLE/COM Object Viewer, which comes with your copy of VC++, on the client machine. Expand the tree view of objects at the "All Objects" branch as seen here in Figure 7.4.

Figure 7.4   OLE/COM Object Viewer

Expand the tree view at the "All Objects" branch.

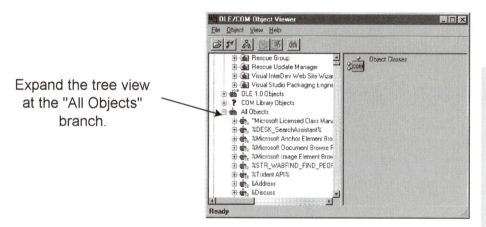

2. Next, locate your COM class in this list (it obviously must have already been registered once by the Studio or by using regsvr32.exe), and click once on it. This will open a raw view of the registry in the right-hand window as seen in Figure 7.5. Click on the "Implementation" tab of this property sheet to view yet another property sheet. Click first on the "Inproc Server" tab and clear the "Path to Implementation" edit box. Do the same for the "Local Server" tab. Then, click on the "Activation" tab of the main property sheet and enter the name of the machine that your server is on. Click on the "Registry" tab to save your settings.

3. Register your EXE server on its machine as usual or with OLE/COM Object Viewer, except this time leave the Activation tab alone and fill in the Local Server path.

4. If you are running Windows NT or Windows 2000 on the server machine, make sure you have an account on that machine. In other words, if you signed into the machine that the client is running on using the account and password "Bill"/"Pass", then you should have the exact same account and password on the server machine. COM just uses that account and password to run that application on that machine.

II

7

Figure 7.5 Registering your COM Server

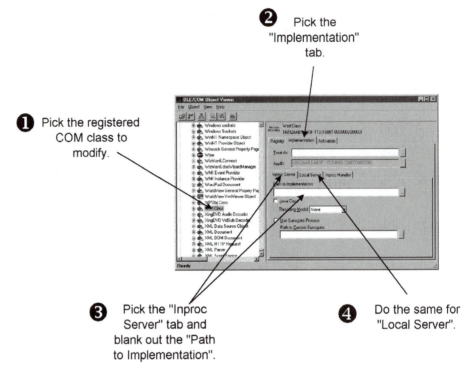

❷ Pick the "Implementation" tab.

❶ Pick the registered COM class to modify.

❸ Pick the "Inproc Server" tab and blank out the "Path to Implementation".

❹ Do the same for "Local Server".

Notes

- Debugging your COM EXE server can be a tedious problem. The debugger will only run one process at a time — which is okay for a dll server that does, in fact, run in the process of the client. However, for an executable, the Studio actually invokes another Studio when you step into the server to run the server. This doesn't allow you to set up any breakpoints in the server. However, what you can do is bring up the other Studio yourself, put your breakpoints on and then execute the Server. Then, bring up the client, which will now attach to your running server rather than start its own. Don't reverse the steps, however, and bring up the client first or it will spawn its own server again and the server in your Studio will be ignored.

- Sometimes when you build your EXE server, the Studio will complain that it can't write to the EXE file due to sharing problems. This means one or more of your servers are still running. And because an EXE server

Example 20   Writing a COM EXE Service Using ATL   **203**

doesn't have an interface, you'll have to kill it using the task list of your operating system (press ctrl-alt-delete to invoke it). Unfortunately, this doesn't always work. On NT/2000 especially, when you go to kill the server, it will respond that you don't have the privilege. To kill a stubborn server like this, just right click on it in the task list to attach a debugger to it. Once it has come up into the Studio's debugger, just click the stop-debugging toolbar button and exit the Studio.

Example 20   Writing a COM EXE Service Using ATL

Objective

You would like to write a COM server that resides in an NT Service application using ATL

Strategy

An NT Service is an application that starts at boot up time and is controlled thorough the Control Panel's "Service's" applet. Windows 9x doesn't support services, however, you can get the same effect by writing an EXE program and sticking it in the `autoexec.bat` file in the root directory. All you miss out on is the ability to pause the application through the Control Panel.

Writing a ATL service is again not unlike a DLL or regular EXE server. However, an ATL service is typically used to also perform some function other than just house one or more COM classes. In fact, the usual scenario is that the service starts a thread that performs some function, such as logging errors or generating reports and the COM object is simply used as a way for any other application to talk to this function (i.e., send it data). We will therefore look at this scenario.

Steps

Create the EXE Project and its COM Classes

1. Create the project using ATL COM AppWizard as shown in Example 18 (page 192). Pick EXE service and notice that the options at the bottom

II

7

of the dialog box are disabled. To find out why, please refer to the last example.

Add a Thread and Event Flags to Your Project

We will now add a thread to our project. The reason we have to add a thread rather than just perform our functionality in the main process is that the main process is tied up with processing the message loop for the service — the one that allows the Control Panel to control the thread. We sue the Win32 API to create our thread which will immediately go into a wait loop waiting for an event. The COM class we create for this service will then be responsible for setting these events from within their methods.

1. At the top of your project's main .cpp file, the one with CServiceModule _Module; defined at the top, add your event flags and thread definitions:

```
// thread stuff
#include <process.h>     /* _beginthread, _endthread */
long glArg;
HANDLE ghNow, ghDie, ghDone, ghThread;
void WzdThread(LPVOID param);
```

2. Down below, just before the service goes into its message loop, create the events and the thread:

```
    // create thread events and start thread
    ghNow = CreateEvent (NULL, TRUE, FALSE, NULL);
    ghDie = CreateEvent (NULL, TRUE, FALSE, NULL);
    ghDone = CreateEvent (NULL, TRUE, FALSE, NULL);
    ghThread = (HANDLE)_beginthread(WzdThread, 0, NULL);

    // the message loop that was put here by the ATL COM AppWizard
    MSG msg;
    while (GetMessage(&msg, 0, 0, 0))
        DispatchMessage(&msg);
```

- The service will leave the message loop when it receives a WM_QUIT message. At this point, you want to tell the thread to also terminate and then deallocate the event flags. Rather than kill the

Example 20   Writing a COM EXE Service Using ATL   **205**

thread, which can be messy, we set an event flag created just for the
occasion that tells the thread to leave its wait loop:

```
// service had terminated, tell thread to die
SetEvent(ghDie);
if (WaitForSingleObject(ghDone, 1000) == WAIT_TIMEOUT)
{
    // thread didn't terminate, kill it...
    DWORD dwCode=0;
    TerminateThread(ghThread, dwCode);
}
CloseHandle(ghDone);
CloseHandle(ghDie);
CloseHandle(ghNow);
```

- Now create the thread that listens to these event flags, perhaps
  times out to perform some chore, but definitely returns when the
  ghDie event is set. Such a thread might be written like this:

```
// WzdThread.cpp: Implementation of WzdThread thread

#include "stdafx.h"
#include <process.h>     /* _beginthread, _endthread */


//////////////////////////////////////////////////////////////////////
   ////////
// event stuff
extern long lArg;
extern HANDLE ghNow, ghDie, ghDone;


// the thread
void WzdThread(LPVOID param)
{
    // thread will work in background
    ::SetThreadPriority(::GetCurrentThread(),THREAD_PRIORITY_BELOW_NORMAL);

    // forever
    while (true)
```

II

7

```
{
    // wait for termination, timeout or a now event
    int rVal;
    HANDLE hEventArray[2]  = {ghDie, ghNow};
    if ((rVal=WaitForMultipleObjects(2, hEventArray, FALSE,
    30000))==WAIT_OBJECT_0)
        break; // terminate now!

    switch (rVal)
    {
////////////////////////////////////////////////////////////////////
    //////
///////////////////////   NOW EVENT ////////////////////////////////
////////////////////////////////////////////////////////////////////
    //////
    case WAIT_OBJECT_0+1:
        ResetEvent(ghNow);


        break;
    }
    }

    // end thread
    SetEvent(ghDone);
    _endthread();
}
```

Add Methods to Your COM Class

1. You can now set or reset these events from your COM class's methods to control this thread. To pass data to the thread, store them in statically-global variables and set the event flag. For thread safety, only store to the variable when it's empty and only read from it within the thread when the event has been set. The thread can then reset the variable to empty when it's done. You can also set up critical sections around any access to shared data.

Example 21   Extending Your ATL COM Class   **207**

Register Your Service

1. Your service project is written to handle three command line flags for
   registering and unregistering your server:

`/service`	registers your server in the system registry *and* as a service
`/unregserver`	not only removes the server from the system registry, but also deletes it as a service
`/regserver`	registers your server as just a plan EXE

Although only NT and 2000 support system services, in a pinch you can
still run your server as a plan executable on a 9x machine.

Notes

- A service is also difficult to debug. So instead of debugging service, register your COM server as a plan EXE server first and debug it there. You can also hook a debugger up to it in Windows NT/2000 in the task list.

Example 21   Extending Your ATL COM Class

Objective

Rather than change an existing COM class interface and perhaps disable
any client that uses it, you would like to add a new interface to it that adds
additional functionality to the first.

Strategy

One of the tenants of COM is that you should never ever change an interface. Of course, in real life within a company, it happens all the time. But if
you want to follow by the rules, we will review how to add a new extension
to an old COM class using ATL. Mostly what is involved is just adding a
new COM class to your existing project using the ATL Object Wizard and
then manually editing both to be aware of each other.

Steps

Edit the Old COM Class

We will be modifying the old COM class's .h file to split up its class into a base class and a derived class (this is to prevent link errors later):

1. Change the class name from, as an example, CWzd to CWzdBase.

2. Create a new class called CWzd at the bottom of the .h file that looks like this:

```
// new class
class CWzd :
     public CWzdBase,
     public CComCoClass<CWzd, &CLSID_Wzd>
{
public:

DECLARE_REGISTRY_RESOURCEID(IDR_WZD)
};
```

3. In the old class definition, comment out the CComCoClass<> in the derivations.

4. In the old class definition, comment out the DECLARE_REGISTRY_RESOURCE ID(IDR_WZD) macro.

5. To get a general idea of what the final class would look like, please refer to the listing at the bottom of this example.

Add the New COM Class

1. Add the new COM class to this server as usual using the ATL Object Wizard. In this example, we name it CWzdEx.

2. Modify the new class's .h file as follows to replace the CComObjectRootEx<> in the derivation to your CWzdBase class.

3. In the interface map, replace COM_INTERFACE_ENTRY(IDispatch) with COM_INTERFACE_ENTRY2(IDispatch, IWzdEx).

4. Add a chaining macro to the bottom of the map to chain this interface with the COM class we're extending:

```
COM_INTERFACE_ENTRY_CHAIN(CWzdBase)
```

Example 21   Extending Your ATL COM Class   **209**

5. Because this new class is derived from the old class, you can now use any or all of the old class's methods in your new class, perhaps even passing an entire method to the base class.

Use the New Class Extension

1. Create the extended COM class as usual using smart pointers:

```
IWzdExPtr pWzdEx;
HRESULT hr=pWzdEx.CreateInstance(__uuidof(WzdEx));
if (FAILED(hr))
{
    _com_error err(hr);
    AfxMessageBox(err.ErrorMessage());
    return;
}
pWzdEx->Method2(4321);
```

2. To access the methods of the class we extended, use smart pointers to query for its interface.

```
IWzdPtr pWzd1(pWzdEx);
pWzd1->Method1(1234);
```

3. Note that you *cannot* go backwards — upon creating the old class, you can't get a pointer to the interface of the new class.

Notes

- Note that even though the new class is derived from the old class, that you cannot access the old class's methods from the new class. This gets into the issue of COM classes and why they can't be derived from one another. For more on why, please refer to Chapter 3 and Example 26 (page 226). If you want, you can simply add the old class's methods to your new class and call the old class's method inside.

CD Notes

- Build the project on the CD and put a breakpoint in the OnTest() button handler and then step through the old and the new implementations.

II

7

Listings — Edited Old Class's .H File

```
// Wzd.h : Declaration of the CWzd

#ifndef __WZD_H_
#define __WZD_H_

#include "resource.h"        // main symbols


/////////////////////////////////////////////////////////////////////
    /////////
// CWzd
class ATL_NO_VTABLE CWzdBase :
    public CComObjectRootEx<CComSingleThreadModel>,
//  public CComCoClass<CWzd, &CLSID_Wzd>, //moved down
    public IDispatchImpl<IWzd, &IID_IWzd, &LIBID_SERVERLib>
{
public:
    CWzdBase()
    {
    }


//DECLARE_REGISTRY_RESOURCEID(IDR_WZD)        //moved down

DECLARE_PROTECT_FINAL_CONSTRUCT()

BEGIN_COM_MAP(CWzdBase)
    COM_INTERFACE_ENTRY(IDispatch)
    COM_INTERFACE_ENTRY(IWzd)
END_COM_MAP()

// IWzd
public:
    STDMETHOD(Method1)(/*[in]*/ long lArg);
};

// new class
```

Example 21   Extending Your ATL COM Class   **211**

```
class CWzd :
    public CWzdBase,
    public CComCoClass<CWzd, &CLSID_Wzd>
{
public:


DECLARE_REGISTRY_RESOURCEID(IDR_WZD)
};



#endif //__WZD_H_
```

Listings — Edited New Class's .H File

```
// WzdEx.h : Declaration of the CWzdEx

#ifndef __WZDEX_H_
#define __WZDEX_H_

#include "resource.h"        // main symbols
#include "Wzd.h"
/////////////////////////////////////////////////////////////////////
    /////////
// CWzdEx
class ATL_NO_VTABLE CWzdEx :
    public CWzdBase,
//   public CComObjectRootEx<CComSingleThreadModel>,
    public CComCoClass<CWzdEx, &CLSID_WzdEx>,
    public IDispatchImpl<IWzdEx, &IID_IWzdEx, &LIBID_SERVERLib>
{
public:
    CWzdEx()
    {
    }

DECLARE_REGISTRY_RESOURCEID(IDR_WZDEX)
```

II

7

```
DECLARE_PROTECT_FINAL_CONSTRUCT()

BEGIN_COM_MAP(CWzdEx)
    COM_INTERFACE_ENTRY2(IDispatch, IWzdEx)
    COM_INTERFACE_ENTRY(IWzdEx)
    COM_INTERFACE_ENTRY_CHAIN(CWzdBase)
END_COM_MAP()

// IWzdEx
public:
    STDMETHOD(Method2)(/*[in]*/long lArg);
};
#endif //__WZDEX_H_
```

Example 22   Writing an ATL Server that Supports Late Binding

Objective

You would like to write your ATL server — DLL or EXE — so that it supports late binding (automation) and therefore can be used by Visual Basic and Visual J++ clients.

Strategy

Actually, we did this in Examples 18 and 19 — you just didn't know it. Unlike MFC, which makes you very aware of what type of interface you're supporting, as long as you don't put any late-binding hostile argument types into your methods, the ATL support classes take care of both interfaces invisible to you.

Steps

Create a Late Binding COM Class

1. Follow all the steps in Example 18 or 19. Just make sure to specify "Dual" when selecting the attributes for your COM class in the ATL Object Wizard. And if the MIDL compiler complains about an argument

Example 23   Writing an ATL Singleton Server   **213**

type not being compatible with OLE Automation, this time you definitely have to change the type to something it will support.

Notes

- Although COM objects are suppose to work with any client, each client has its very own selection of argument types you should be working with. Please see Chapter 2 and Examples 31, 32, and 33 (beginning on page 246) for more.

Example 23   Writing an ATL Singleton Server

Objective

You would like to write an ATL COM server — DLL or EXE — which will only create one instance of itself no matter how many times one or more clients ask for a new instance of it.

Strategy

A class that only creates one instance of itself is called a singleton. As we saw in the last chapter, it's not very trivial to create a singleton using MFC. In ATL, it's almost too easy — you just add a macro to the COM class's .h file.

Steps

Make a COM Class a Singleton

1. To make a COM class into a singleton just add the following macro to its .h file just about the DECLARE_REGISTRY_RESOURCEID macro:

```
DECLARE_CLASSFACTORY_SINGLETON(CWzd)
```

where CWzd is the name of the class.

Notes

- You've probably noticed by now if you're reading this chapter from start to finish, that a lot of the configuration of an ATL COM class is done in its .h file leaving just method implementations in its .cpp file.

CD Notes

- When executing the project on the accompanying CD, you will notice that the same object pointer is returned every time `CreateInstance()` is called.

Example 24   Writing an ATL COM Tearoff Server

Objective

You would like to COM class that contains tearoff interfaces.

Strategy

A tearoff interface gives you a way to create only a portion of your COM class at any one time. Why would you need that kind of functionality? Usually, it's in the situation where you have a COM class that has lots of functionality but it's only infrequently or optionally used based on the hardware you have or some other factor.

To add these interfaces, we will be doing lots of manual editing to the IDL and .h files of our COM "base" class and making liberal use of ATL's helpful macros.

Steps

Create the COM Server and Class

1. Create the ATL server and COM class as usual using the wizards as seen in Example 18 (page 192).

Add the Tearoff Interface to the IDL File

1. In the IDL file of the newly created COM class, manually copy and paste the existing COM class's interface for as many tearoff interfaces you need. We use this empty interface just so we don't have to write one from scratch. However, make sure to change the GUID for each interface you add — which you can do by just adding 20 to the first group of numbers. Also add these interfaces to the Library definition below so that they become part of the type library.

Example 24   Writing an ATL COM Tearoff Server   **215**

2. Please see an example IDL file below.

Add the Tearoff Interface to the .H File

1. In the .h file, copy and paste the original COM class definition for each tearoff interface you're adding. In the new tearoff classes, change the base class(es) to CComTearOffObjectBase<CWzd> and remove everything else except the interface map as in this example:

```
class ATL_NO_VTABLE CWzdTear :
    public IDispatchImpl<IWzdTear, &IID_IWzdTear, &LIBID_SERVERLib>,
    public CComTearOffObjectBase<CWzd>
{
public:
    CWzdTear()
    {
    }


BEGIN_COM_MAP(CWzdTear)
    COM_INTERFACE_ENTRY(IWzdTear)
END_COM_MAP()

public:
};
```

2. In the original COM class's interface map, add a COM_INTERFACE_ENTRY_TEAR_OFF macro for each tearoff interface:

```
    COM_INTERFACE_ENTRY_TEAR_OFF(IID_IWzdTear, CWzdTear)
```

3. Don't worry, you don't have to edit the .cpp file too. After you've saved your work, you can use the ClassView tab of the Workspace View to add methods and properties to tearoff interfaces.

4. Please see an example .H file below under Listings.

Use the Tearoff Interface

1. You can now create this COM class like any other:

```
IWzdPtr pPtr;
HRESULT hr=pPtr.CreateInstance(__uuidof(Wzd));
if (FAILED(hr))
{
    _com_error err(hr);
    AfxMessageBox(err.ErrorMessage());
    return;
}
pPtr->Method1(1234);
```

2. To create one of the tearoff interfaces using smart pointers, use the following:

```
IWzdTearPtr pTear(pPtr);
pTear->Method2(1234);
```

3. When the smart pointer releases this object, the object is destroyed immediately!

Notes

- Only use a tearoff interface if the classes are related. If you just have a whole bunch of unrelated and seldom used COM classes, just leave them in separate files.

CD Notes

- When executing the project on the accompanying CD, you will notice that the tearoff interface is created using QueryInterface() and destroyed immediately when the object is released.

Example 24   Writing an ATL COM Tearoff Server   **217**

Listings — Example IDL File

```
// Server.idl : IDL source for Server.dll
//

// This file will be processed by the MIDL tool to
// produce the type library (Server.tlb) and marshalling code.


////////////////////////////////////////////////////////
////////////////////////////////////////////////////////

import "oaidl.idl";
import "ocidl.idl";
    [
        object,
        uuid(C177116E-9AAA-11D3-805D-000000000000),
        dual,
        helpstring("IWzd Interface"),
        pointer_default(unique)
    ]
    interface IWzd : IDispatch
    {
        [id(1), helpstring("method Method1")] HRESULT Method1([in]
    long lArg);
    };
    [
        object,
        uuid(C177118E-9AAA-11D3-805D-000000000000),
        dual,
        helpstring("IWzdTear Interface"),
        pointer_default(unique)
    ]
    interface IWzdTear : IDispatch
    {
        [id(1), helpstring("method Method2")] HRESULT Method2([in]
    long lArg);
```

```
    };
[
    uuid(C1771162-9AAA-11D3-805D-000000000000),
    version(1.0),
    helpstring("Server 1.0 Type Library")
]
library SERVERLib
{
    importlib("stdole32.tlb");
    importlib("stdole2.tlb");


    [
        uuid(C177116F-9AAA-11D3-805D-000000000000),
        helpstring("Wzd Class")
    ]
    coclass Wzd
    {
        [default] interface IWzd;
        interface IWzdTear;
    };
};
```

Listings — Example .H File

```
// Wzd.h : Declaration of the CWzd

#ifndef __WZD_H_
#define __WZD_H_

#include "resource.h"       // main symbols


///////////////////////////////////////////////////////////////////
    ////////
// CWzdTear
class CWzd;
class ATL_NO_VTABLE CWzdTear :
    public IDispatchImpl<IWzdTear, &IID_IWzdTear, &LIBID_SERVERLib>,
```

Example 24   Writing an ATL COM Tearoff Server   **219**

```
    public CComTearOffObjectBase<CWzd>
{
public:
    CWzdTear()
    {
    }

//DECLARE_REGISTRY_RESOURCEID(IDR_WZD)

//DECLARE_PROTECT_FINAL_CONSTRUCT()

BEGIN_COM_MAP(CWzdTear)
    COM_INTERFACE_ENTRY(IWzdTear)
//  COM_INTERFACE_ENTRY(IDispatch)
END_COM_MAP()

// IWzd
public:
    STDMETHOD(Method2)(long lArg);
};

/////////////////////////////////////////////////////////////////
    ////////
// CWzd
class ATL_NO_VTABLE CWzd :
    public CComObjectRootEx<CComSingleThreadModel>,
    public CComCoClass<CWzd, &CLSID_Wzd>,
    public IDispatchImpl<IWzd, &IID_IWzd, &LIBID_SERVERLib>
{
public:
    CWzd()
    {
    }

DECLARE_REGISTRY_RESOURCEID(IDR_WZD)

DECLARE_PROTECT_FINAL_CONSTRUCT()
```

```
BEGIN_COM_MAP(CWzd)
    COM_INTERFACE_ENTRY(IWzd)
    COM_INTERFACE_ENTRY(IDispatch)
    COM_INTERFACE_ENTRY_TEAR_OFF(IID_IWzdTear, CWzdTear)
END_COM_MAP()

// IWzd
public:
    STDMETHOD(Method1)(long lArg);
};
#endif //__WZD_H_
```

Example 25   Writing an ATL COM Server that has a Connection Point

Objective

You would like to write a COM server — DLL or EXE — that supports a connection point.

Strategy

Not all communication between client and server is one-way. Sometimes the client might want the server to call it when some event occurs. To do this in COM, the client temporarily becomes a server itself implementing its own COM object with a method that the server calls. The client then registers a pointer to this object with the server for the server to call. In the parlance of COM, a server that allows a client to register a "callback" object with it, is said to have a "connection point".

Just as a COM server itself might have to support an early or late-binding interface, so must a connection point. Therefore, first we will show how to add a connection point that supports late binding for Visual Basic and Visual J++ clients using just the ClassView. After that, we'll review how to add a connection point that supports early binding for MFC clients, which atypically for ATL requires a lot of manual editing using the ATL Object Wizard, the Connection Point map, and a function you have to write yourself.

Example 25   Writing an ATL COM Server that has a Connection Point   **221**

Steps

Add a Late-Binding Connection Point

When a sink interface is added to an MFC client, you have to create an Interface Project to define just what methods the server can call in the client (as shown in Example 9, page 143). But you can't add an IDL file to a VB application. So how can VB not only know the name of the interface method to call but also the type of the calling arguments? Because you define it in the COM server's IDL file and VB or VJ++ accesses this definition through the server's type library.

1. Create the COM Class as usual using the ATL Object Wizard, however make sure to select the "Support Connection Points" property. This causes the wizard to add a Connection Point map to the .h file of your class and to also add an event interface to your IDL file that will be defining the callback method(s) in the VB client.

2. Open the ClassView and right-click on this event interface and add the definition of the method or methods that your VB or VJ++ client will be supporting.

3. A new class will be created for your COM server project that will contain the method(s) you added in the last step, although each name will be prefixed with "Fire_". You can now call this method(s) from anywhere in this ATL COM class.

Note: If you look in the IDL file, you'll notice that ATL only adds this event interface to the type library and uses the "source" attribute in the coclass declaration. The "source" attribute tells the MIDL compiler that whatever COM server implements this interface doesn't have to implement this particular class. Instead, this definition resides in the type library for VB to access when implementing the sink in the client.

II

7

Add an Early-Binding Connection Point

You can add an IDL file to an MFC application to implement an early-binding sink interface (see Example 14, page 163). So, here we will see how to add a connection point that it can connect to.

1. Create the COM Class as usual using the ATL Object Wizard, however make sure to select the "Support Connection Points" property. This causes the wizard to add a Connection Point map to the .h file of your class. It also adds a late-binding interface for events that we will ignore.

2. Somehow include the Interface ID of the client's sink COM class in this .h file (please see Example 9, page 143, on how to create this class). In other examples, we've included guids by copying them from the interface server project and putting them in their own guids.h file like so:

```
#if !defined guids_h
#define guids_h

const IID IID_IWzdSink =
    {0x20050FE0,0xA719,0x11d3,{0xA3,0x98,0x00,0xC0,0x4F,0x57,0x0E,
    0x2C}};

#endif
```

3. Locate the Connection Point map that was created in your ATL COM class's generated .h file and add a macro that will reference the client's Interface ID:

```
BEGIN_CONNECTION_POINT_MAP(CWzd)
    CONNECTION_POINT_ENTRY(IID_IWzdSink)
END_CONNECTION_POINT_MAP()
```

4. At the top of this .h file, you will also need to add a class as seen below. This class contains one method that can be called by any method of your COM class and will loop through all clients that have registered themselves with your class:

```
template <class T>
class CProxyIWzdSinkEvents : public IConnectionPointImpl<T,
    &IID_IWzdSink, CComDynamicUnkArray>
{
public:
```

Example 25   Writing an ATL COM Server that has a Connection Point   **223**

```
    HRESULT Fire_WzdSink( long lArg )
    {
        T* pT = (T*)this;
        pT->Lock();
        HRESULT ret;
        IUnknown** pp = m_vec.begin();
        while (pp < m_vec.end())
        {
            if (*pp != NULL)
            {
                IWzdSink *pWzdSinks =
    reinterpret_cast<IWzdSink*>(*pp);
                ret = pWzdSinks->Callback( lArg );
            }
            pp++;
        }
        pT->Unlock();
        return ret;
    }
};
```

5. In this class, change the name of the method to whatever is appropriate and also change the name of the class and method it calls to whatever class and method the client has registered with your server. The number of arguments and types can also be changed.

6. Add this new class as one of the base classes of the COM class that the wizard created for you:

```
class ATL_NO_VTABLE CWzd :
    public CComObjectRootEx<CComSingleThreadModel>,
    public CComCoClass<CWzd, &CLSID_Wzd>,
    public IConnectionPointContainerImpl<CWzd>,
    public IDispatchImpl<IWzd, &IID_IWzd, &LIBID_SERVERLib>,
    public CProxyIWzdSinkEvents<CWzd> //<<<<<<added
```

7. You can now call your client in any of your COM class's methods with:

```
    HRESULT hr=Fire_WzdSink(1234);
```

8. You may notice that the code you just added manually was added automatically for you when creating a late-binding connection point. For a complete listing of an example COM class that supports an early-binding connection point, please refer to the listing at the end of this example.

Notes

- To connect to this class using VB or J++, see Chapter 8.
- The return trip from server back to client must follow all the same rules as any COM object when implementing an early-binding connection point. As an example, if you use argument types that aren't supported by the COM dll and you're going across process boundaries, you must create and register the proxy/stub dll as with any COM class. As usual, late-binding interfaces take care of themselves.

CD Notes

- Please refer to the CD Notes under the clients that implement sinks in Chapter 8.

Listings — Tooltip Class

```cpp
// Wzd.h : Declaration of the CWzd

#ifndef __WZD_H_
#define __WZD_H_

#include "resource.h"        // main symbols
#include "..\ISinkSrv\IWzdSink.h"
EXTERN_C const IID IID_IWzdSink;


/////////////////////////////////////////////////////////////////////
    ////////
template <class T>
class CProxyIWzdSinkEvents : public IConnectionPointImpl<T,
    &IID_IWzdSink, CComDynamicUnkArray>
{
public:
    HRESULT Fire_WzdSink( long lArg )
```

Example 25   Writing an ATL COM Server that has a Connection Point   **225**

```
    {
        T* pT = (T*)this;
        pT->Lock();
        HRESULT ret;
        IUnknown** pp = m_vec.begin();
        while (pp < m_vec.end())
        {
            if (*pp != NULL)
            {
                IWzdSink *pWzdSinks =
        reinterpret_cast<IWzdSink*>(*pp);
                ret = pWzdSinks->Callback( lArg );
            }
            pp++;
        }
        pT->Unlock();
        return ret;
    }
};



///////////////////////////////////////////////////////////////////////
    ////////
// CWzd
class ATL_NO_VTABLE CWzd :
    public CComObjectRootEx<CComSingleThreadModel>,
    public CComCoClass<CWzd, &CLSID_Wzd>,
    public IConnectionPointContainerImpl<CWzd>,
    public IDispatchImpl<IWzd, &IID_IWzd, &LIBID_SERVERLib>,
    public CProxyIWzdSinkEvents<CWzd>
{
public:
    CWzd()
    {
    }

DECLARE_REGISTRY_RESOURCEID(IDR_WZD)
```

II

7

```
DECLARE_PROTECT_FINAL_CONSTRUCT()

BEGIN_COM_MAP(CWzd)
    COM_INTERFACE_ENTRY(IWzd)
    COM_INTERFACE_ENTRY(IDispatch)
    COM_INTERFACE_ENTRY(IConnectionPointContainer)
END_COM_MAP()
BEGIN_CONNECTION_POINT_MAP(CWzd)
    CONNECTION_POINT_ENTRY(IID_IWzdSink)
END_CONNECTION_POINT_MAP()


// IWzd
public:
    STDMETHOD(Method1)(/*[in]*/ long lArg);
};

#endif //__WZD_H_
```

Example 26   Aggregating a COM Object Using ATL

Objective

When writing an ATL COM Server, you would like to derive some of its functionality from another COM object.

Strategy

As mentioned in the last chapter and in Chapter 3, aggregation is COM's weak version of class inheritance. Your COM object can't access the "base" class's functionality as easily as in the classic sense, but it can use Query-Interface() to get a pointer to the other object and access its functionality that way.

As seen in the last chapter with MFC, aggregation is a matter of the "derived" class automatically creating and releasing the "base" class.

Example 26   Aggregating a COM Object Using ATL   **227**

But you'll find even with ATL's ubiquitous macros that this job isn't any easier than it was with MFC.

Steps

Create the "Base" COM Class

1. When using the ATL Object Wizard to create the COM class you'll be aggregating, make sure to specify "Yes" (the default) or "Only" when picking an Aggregation option.

Create the "Derived" COM Class

1. You will need to access the Interface and Class IDs of the "base" COM class. As in prior examples, you can do this by creating a guids.h include file:

```
#if !defined guids_h
#define guids_h

const CLSID CLSID_WzdAggSvr =
    {0x853B853F,0xA88A,0x11D3,{0xA3,0x98,0x00,0xC0,0x4F,0x57,0x0E,
    0x2C}};
const IID IID_IWzdAggSvr =
    {0x853B853E,0xA88A,0x11D3,{0xA3,0x98,0x00,0xC0,0x4F,0x57,0x0E,
    0x2C}};

#endif
```

2. Use the ATL Object Wizard to create the class as usual.
3. At the bottom of the generated .h file, add a new member variable that will contain a pointer to the aggregated object:

```
private:
    LPUNKNOWN m_pUnknown;
```

4. Initialize this member variable to NULL in the constructor:

```
CWzdSrv()
{
    m_pUnknown=NULL;
}
```

5. Add a final constructor to the class where you will create an instance of the aggregated object:

```
HRESULT FinalConstruct( )
{
    HRESULT hr=::CoCreateInstance(CLSID_WzdAggSvr,
                GetControllingUnknown(), CLSCTX_ALL,
                IID_IUnknown, (LPVOID*)&m_pUnknown);
    return hr;
}
```

6. Add a final release to the class where you will release the aggregated object:

```
void FinalRelease()
{
    m_pUnknown->Release();
}
```

7. Add the following macro which allows you to call `GetControllingUnknown()` in step 5:

```
DECLARE_GET_CONTROLLING_UNKNOWN()
```

8. Add an aggregating macro to the interface map specifying the interface that's being aggregated and the instance of the object that was aggregated:

```
BEGIN_COM_MAP(CWzdSrv)
    COM_INTERFACE_ENTRY(IWzdSrv)
    COM_INTERFACE_ENTRY(IDispatch)
    COM_INTERFACE_ENTRY_AGGREGATE(IID_IWzdAggSvr,m_pUnknown)
END_COM_MAP()
```

Example 26   Aggregating a COM Object Using ATL   **229**

Note: You can also use the following aggregation macro and avoid step 5 altogether:

```
COM_INTERFACE_ENTRY_AGGREGATE_AUTO(CLSID_WzdAggSvr,IID_IWzdAggSvr,
  m_pUnknown)
```

9. For an example listing of the aggregating class's .h file, please refer to the end of this example.

Access the Aggregated Class Using the COM API

1. Create the "derived" second COM class as usual using ::CoCreateInstance() as seen in Example 1 (page 104).

2. To access the "base" class, you can't simply call its methods using the same class pointer. Instead, you will need to use that pointer's QueryInterface() method to get a pointer to that class like so:

```
IWzdAgg *iWzdAgg=NULL;
hr=iWzd->QueryInterface(IID_IWzdAgg, (LPVOID*)&iWzdAgg);
if (FAILED(hr))
{
    _com_error err(hr);
    AfxMessageBox(err.ErrorMessage());
    return;
}
```

where iWzd is the pointer to the "derived" class and iWzdAgg is a pointer to the "base" class.

3. You can then access the methods of these two classes as before:

```
iWzdAgg->Method1(1234,&lArg);
iWzd->Method2(lArg2, ulArg);
```

4. You must still release both objects — aggregation doesn't even take that burden away from you:

```
iWzd->Release();
iWzdAgg->Release(); // must still release aggregated interface
```

Notes

- Please refer to Example 16 (page 173) for other notes on aggregation.

CD Notes

- Open the `test.dsw` project and build the debug version. Place a break-point in the `OnTest()` button handler and run the application. Then, step through as both the aggregated and aggregating COM objects are created and used.

Listings — Tooltip Class

```
// WzdSrv.h : Declaration of the CWzdSrv

#ifndef __WZDSRV_H_
#define __WZDSRV_H_


#include "resource.h"        // main symbols


#include "guids.h"
/////////////////////////////////////////////////////////////////////////
/////////
// CWzdSrv
class ATL_NO_VTABLE CWzdSrv :
    public CComObjectRootEx<CComSingleThreadModel>,
    public CComCoClass<CWzdSrv, &CLSID_WzdSrv>,
    public IDispatchImpl<IWzdSrv, &IID_IWzdSrv, &LIBID_SERVERLib>
{
public:
    CWzdSrv()
    {
        m_pUnknown=NULL;
    }


    HRESULT FinalConstruct( )
    {
        HRESULT hr=::CoCreateInstance(CLSID_WzdAggSvr,
```

Example 26   Aggregating a COM Object Using ATL   **231**

```
                            GetControllingUnknown(), CLSCTX_ALL,
                            IID_IUnknown, (LPVOID*)&m_pUnknown);
    return hr;
}


void FinalRelease()
{
    m_pUnknown->Release();
}

DECLARE_GET_CONTROLLING_UNKNOWN()

DECLARE_REGISTRY_RESOURCEID(IDR_WZDSRV)

DECLARE_PROTECT_FINAL_CONSTRUCT()

BEGIN_COM_MAP(CWzdSrv)
    COM_INTERFACE_ENTRY(IWzdSrv)
    COM_INTERFACE_ENTRY(IDispatch)
    COM_INTERFACE_ENTRY_AGGREGATE(IID_IWzdAggSvr,m_pUnknown)
END_COM_MAP()

// IWzdSrv
public:
    STDMETHOD(Method2)(/*[in]*/ long lArg, /*[in]*/ unsigned
    long ulArg);

private:
    LPUNKNOWN m_pUnknown;
};

#endif //__WZDSRV_H_
```

II

7

Chapter 8

Writing COM Servers with VB and VJ++

We've seen how difficult it can be to write a COM server using ATL or MFC — especially MFC. We will now discover how easy it is to create a server using Visual Basic or Visual J++. In most cases, the fact that you're creating a COM server at all is hidden from you — no IDL file to tweak, no COM rules to remember. And both of these languages create servers that are much more robust than the equivalent C++ code where you can get into all sorts of trouble. But of course, this simplicity and robustness comes at the price of a performance hit and a lack of choice. For performance, you can expect your VB or VJ++ COM server to run 40% slower than your C++ server. And for lack of choice, you still can't write a VB or VJ++ server that will support MTS/COM+ transactions. But if you just don't have the people or the time…

The examples in this chapter include:

Example 27 Writing a COM ActiveX Server Using Visual Basic
where we create an ActiveX server using VB.

Example 28 Adding a Sink to a Visual Basic Client where we connect our VB client to an ATL COM server's connection point using late binding.

Example 29 Writing a COM DLL Server Using Visual J++ where we create a J++ COM server.

Example 30 Adding a Sink to a Visual J++ Client where we connect our VJ++ client to an ATL COM server's connection point using late binding.

Example 27 Writing a COM ActiveX Server Using Visual Basic

Objective

You would like to write a COM server using Visual Basic.

Strategy

Visual Basic is fully automated to create a COM server for you using the ActiveX standard. You use the Studio to create the project for DLL or EXE or a DLL that supports a control. You use the Visual Basic Class Builder to add the COM classes, methods and events.

Steps

Create the Project

1. In the Visual Basic Studio, click on the "File" and "New" menu items to open the New Project dialog box. Then, pick ActiveX EXE, DLL, or Control.

Add the COM Classes, Methods, and Events

1. Again in the Studio, click on the "Project" and "Add Class Module" menu items to open the "Add Class Module" dialog box. Then, click on "VB Class Builder" to open the VB Class Builder. You'll get a warning

Example 27   Writing a COM ActiveX Server Using Visual Basic   **235**

that the first class added to your project won't work with the VB Class Builder — just click "OK".

2. In the VB Class Builder, right-click on the new class and add methods, properties, events, other classes, etc. as shown in Figure 8.1.

Figure 8.1   Adding Methods with the VB Class Builder

3. When creating a method, you will be allowed to pick any VB argument type because they're a subset of what's supported by COM.

4. You can create ActiveX events to report back to your client. Your client therefore must use an event sink as shown in Example 17 (page 184) rather than the event interface used with non-ActiveX servers.

5. When creating a new class, you will be allowed to "base" one class on another. Unfortunately, this isn't the classic sense of class derivation. Instead, the VB Class Builder simply creates an instance of the "base" class in the "derived" class for you to use there.

6. To save your changes to your classes, you must click on the VB Class Builder's "File" and "Update Project" menu items — otherwise your changes will be lost.

Build the Server

1. Build the project as with any VB project using "File" and "Make *xxx*", where *xxx* is the project name. Building your COM server automatically registers it. On another machine, you can still use regsvr32.

Notes

- The collateral advantage to creating COM classes from VB is that it adds object-oriented programming to the Basic lexicon.

CD Notes

- To test the project on the CD, start the server first and tell it to start when some client has created an instance of it. Then, start the client and step through the server.

Example 28   Adding a Sink to a Visual Basic Client

Objective

You would like to add a sink to your Visual Basic application so that it can connect to a COM class's Connection Point.

Strategy

Connection points and sinks are how non-ActiveX COM servers inform a client when some event happens. Although COM servers written in VB can't themselves have a connection point (you can only create an ActiveX COM server with VB), you do have the ability to "listen" to some other server's connection point using VB's `WithEvents` attribute.

Steps

Add a Sink to Your VB Application

1. First, add the COM class to your VB project as usual using "Project/References..."
2. Then, declare this COM class globally (above all subroutines) as follows:

```
Dim WithEvents IWzdEvent As SERVERLib.Wzd
```

Example 28   Adding a Sink to a Visual Basic Client   **237**

3. Next, add a `Form_Load()` handler to your form by locating "Form" in the Object combo box and "Load" in the Procedure combo box in the form editor. There, create an instance of this COM class like so:

```
Private Sub Form_Load()
    ' connects with object's connection point
    Set IWzdEvent = New SERVERLib.Wzd
End Sub
```

4. Add the actual sink function (the function that the COM server will be calling) by locating "(General)" in the Object combo box and the handler name in the Procedure combo box. The handler name will be the variable name you assigned the class and the callback's method name. The arguments are automatically determined from the type library:

```
Private Sub IWzdEvent_Callback(ByVal lArg As Long)
    MsgBox "sink called"
End Sub
```

5. For a full listing of this, please refer to the end of this example.

Notes

- Because VB can only use COM objects that support late binding, this sink also requires a COM server that supports a late-binding connection point. To create an ATL connection point that supports late binding, please refer to Example 25 (page 220).

CD Notes

- Create and register the server project first using the Developer's Studio. Then, bring the client up into the VB Studio and put a breakpoint on the `IWzdEvent_Callback` function and click "Test" and notice that your VB application was notified of an event by the COM server.

Listings — Visual Basic Listing with Sink

```
Dim WithEvents IWzdEvent As SERVERLib.Wzd
Private Sub Form_Load()
    ' connects with object's connection point
    Set IWzdEvent = New SERVERLib.Wzd
End Sub
Private Sub Command1_Click()
    ' use Call to call COM object methods
    Call IWzdEvent.Method1(1234)


    MsgBox "done"


End Sub
Private Sub Command2_Click()
    End
End Sub
Private Sub IWzdEvent_Callback(ByVal lArg As Long)
    MsgBox "sink called"
End Sub
```

Example 29   Writing a COM DLL Server Using Visual J++

Objective

You would like to write a COM DLL server using Visual J++.

Strategy

Just as with Visual Basic, the VJ++ Developer Studio makes it easy to create a COM DLL server. We will be creating the project using the Studio's "File" menu command and we will be adding methods using the Class Outline view.

Example 29   Writing a COM DLL Server Using Visual J++   **239**

Steps

Create the COM DLL

1. Click on the "File" then "New Projects" to open the "New Project" dialog box. Click on the "New" tab and expand the tree view at the "Visual J++ Projects" branch. Then, click on "Components". Pick the "COM DLL" in the right list control and click "Open". The Studio will now create your COM DLL project and add a class to it. All of the public methods of this class will automatically be exposed to a COM client that creates this server.

2. To easily add methods to the created class, expand the project in the Project Explorer and double-click on the .java file. A Class Outline will then appear in the left view which will contain a tree view that will list this class's name as one of its branches. Right-click on this name to open a menu that will allow you to automatically add methods to your class as shown in Figure 8.2.

Figure 8.2   Adding Methods with the Class Outline

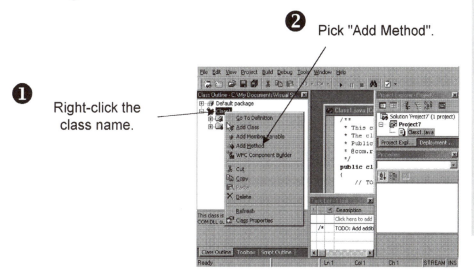

❷ Pick "Add Method".

❶ Right-click the class name.

3. Build the project as usual which will automatically register the DLL. Use regsvr32.exe or the OLE/COM Object Viewer as usual to register the DLL on another machine.

Notes

- As per usual, creating this COM server was a snap but at the price of performance. This late-bound object will typically run 40% slower than the equivalent COM server written in C++ with ATL.

Example 30   Adding a Sink to a Visual J++ Client

Objective

You would like to add a sink to your Visual J++ application so that it can connect to a COM class's Connection Point.

Strategy

Connection points and sinks are how non-ActiveX COM servers inform a client when some event happens. To add a sink to a J++ client, we start by wrapping the COM server in a J++ class. Because you can't stick an IDL file in a J++ project, we get our sink interface from that class wrapper (which in turn, got it from the COM server's type library). We then use Java's "implement" syntax to implement that interface in our client. And finally we use a special Java/VJ++ helper class called `ConnectionPointCookie` to take it from there.

Steps

Add a Sink to Your VJ++ Application

1. Include the COM package in your VJ++ class:

```
import com.ms.com.*;
```

2. Wrap the COM server class with the connection point as usual using the "Project/Add COM Wrapper" menu commands.

3. One of the items extracted when VJ++ was wrapping the COM server was the interface for sink we must implement. Add this implementation

Example 30   Adding a Sink to a Visual J++ Client   **241**

syntax to your J++ class like so (examine the generated wrapper to get the exact name of the interface to implement):

```
public class Form1 extends Form implements server._IWzdEvents
{
    :    :    :
```

4. Implement the method(s) of the sink interface:

```
    public void Callback(int lArg)
    {
    }
```

5. Create a global `ConnectionPointCookie` class variable:

```
public class Form1 extends Form implements server._IWzdEvents
{
   ConnectionPointCookie cookie; // Cookie that connects your event
    public Form1()
    {
        :    :    :
```

6. At some point, create the COM server as usual and then create an instance of the `ConnectionPointCookie` class. You can't do this backwards because the `ConnectionPointCookie` class needs to refer to the COM server:

```
    // create server
    server.Wzd wzd=new server.Wzd();
    // connect to server's connection point
    cookie = new ConnectionPointCookie(
        wzd,                    //source of callbacks
        this,                   //destination of callbacks
        server._IWzdEvents.class); //interface class
```

7. The COM server can now call this J++ class's "`Callback()`" method.

8. To see this class in its entirety, please refer to the end of this example.

Notes

- This example will only work with a COM server that supports late binding, not only for regular method calls, but also for its connection point. As seen in Example 25 (page 220), you can easily add a late-bound connection point to an ATL COM server. I'm afraid I've left you on your own for MFC.

CD Notes

- When executing the project on the accompanying CD, you will notice that the "Callback" method is called right after "wzd.Method1" is called.

Listings — J++ Class with Sink

```java
import com.ms.wfc.app.*;
import com.ms.wfc.core.*;
import com.ms.wfc.ui.*;
import com.ms.wfc.html.*;
import com.ms.com.*;

/**
 * This class can take a variable number of parameters on the command
 * line. Program execution begins with the main() method. The class
 * constructor is not invoked unless an object of type 'Form1' is
 * created in the main() method.
 */

public class Form1 extends Form implements server._IWzdEvents
{
    ConnectionPointCookie cookie; // Cookie that connects your event
    public Form1()
    {
        // Required for Visual J++ Form Designer support
        initForm();

        // TODO: Add any constructor code after initForm call
    }

    /**
     * Form1 overrides dispose so it can clean up the
     * component list.
     */
```

Example 30   Adding a Sink to a Visual J++ Client   **243**

```
public void dispose()
{
    super.dispose();
    components.dispose();
}

private void button1_click(Object source, Event e)
{
    // creater server
    server.Wzd wzd=new server.Wzd();

    // connect to server's connection point
    cookie = new ConnectionPointCookie(
        wzd,                     //source of callbacks
        this,                    //destination of callbacks
        server._IWzdEvents.class);  //interface class

    // call server method
    wzd.Method1(1234);

}

private void Dismiss_click(Object source, Event e)
{
    System.exit(0);
}

public void Callback(int lArg)
{
    int i=0;
}
   :        :    :
}
```

II

8

Chapter 9

COM Communications

In a perfect world, the communication that goes on between a client and its COM server would be no different than any class method call, and a client written in any language could freely exchange its data in its native format with a COM server written in any other language. In reality, COM does take care of the brunt of the work it used to take to communicate between applications and machines. However, it depends on you to carefully configure your IDL file to tell it what to do. And if you go outside of its repertoire of data types it knows how to send and receive, you also need to provide it with your own proxy/stub DLL to help it out. It also can't really get over the hurdle of transparently converting a data type that's native to one language if it just doesn't have an exact equivalent in another.

The examples in this chapter include:

Example 31   Passing Data to a COM Object Using C++   where   we review every way possible to send data between C++ client and server.

Example 32   Passing Interface Pointers Between Threads Using C++   where we look at a set of COM API calls that help ensure that a COM interface pointer stays thread safe, even when it's passed as an argument to another thread.

Example 33   Passing Data Between C++ and Visual Basic   where we look at the limitations that not only late binding imposes on what we can send, but also on VB's own limitations on what it will accept.

Example 34   Passing Data Collections Between Visual C++ & Visual Basic   where we look at an argument type — a server really — which is unique to VB and is used to send data collections between VC++ COM server and VB client.

Example 35   Passing Data Between C++ and Visual J++   where we find that although we aren't as limited by what VJ++ will accept as a data type, but we are still limited by what COM will support with late binding.

Example 31   Passing Data to a COM Object Using C++

Objective

You would like to pass data to and from a COM server written in C++.

Strategy

In this example, we will simply be reviewing the different dances you must perform in order to get your data to your COM server, from IDL syntax to argument types to how to extract the data at both ends.

Steps

Pass Simple Variable Data

To pass chars, shorts, unsigned shorts, longs, unsigned longs, enumerations, floats, and doubles:

1. In the IDL file, use:

```
// enumerator (above any interface definitions)
typedef enum {Monday=2, Tuesday, Wednesday, Thursday, Friday} workday;
    :    :    :
    method([in] unsigned char cArg, [in] short sArg, [in] unsigned
    short usArg, [in] workday enumArg, [in] long lArg, [in]
    unsigned long ulArg,[in] float fArg, [in] double dArg);
```

Example 31   Passing Data to a COM Object Using C++   **247**

2. In the client, use these argument types:

```
char cArg='A'; //assumed to be unsigned
short sArg=1234;
unsigned short usArg=40000;
workday enumArg=Wednesday;
long lArg=12345678L;
unsigned long ulArg=4026531840L;
float fArg=1234.0f;
double dArg=12345678.0;
wzd->method(cArg, sArg, usArg, enumArg, lArg, ulArg, fArg, dArg);
```

3. In the COM server, use: **II**

```
method(unsigned char cArg, short sArg, unsigned short usArg, workday
enumArg,long lArg, unsigned long ulArg, float fArg, double dArg)
```

Note: To use this section, just find the argument type you need and its different incarnation between `.idl`, `.cpp`, and `.h` files.

Pass Simple Variable Data Not Supported By VB or VJ++

The following variable types will not work when talking to a VB or VJ++ client and the MIDL compiler will warn you when you use one of these.

1. In the IDL file for `in`'s and `out`'s, use:

```
[in] boolean bArg, [in] byte byArg, [in] hyper yArg, [in]
unsigned hyper uyArg
```

where `hyper` is a 64 bit integer.

2. In client, use:

```
bool bArg=TRUE;
BYTE byArg=12;
hyper yArg=1234567890123456;  //(int64)
unsigned hyper uyArg=1234567890123;

wzd->method(bArg, byArg, yArg, uyArg);
```

3. In the server, use these argument types:

```
boolean bArg, byte byArg, hyper yArg, unsigned hyper uyArg
```

Pass and Receive Data with [in] and [out]

Until now, we've assumed we were only passing data from the client to the server. We will now adjust our argument attributes to both send and receive data. The "in" attribute means we are sending it to the server. "Out" means we are getting something back from the server in that argument. And "in,out" means we are sending something to that server that will be changed and returned in the same argument. The reason why we need to differentiate what to do with these arguments is so that COM can optimize its communication between server and client so that only the data that needs to be sent is sent.

1. In the IDL file for in's and out's and the use of pointers, use:

```
[in] long lArg1,[out] long *lArg2,[in,out] long *lArg3
```

2. In client, use:

```
    long lArg1=123;// 123 will be passed to method
// but lArg1 will not receive anything back
    long lArg2=456; // 456 will be not be passed to method,
// but will be overwritten by method on return
    long lArg3=789; // 789 will be passed to method
// and will also be overwritten by value returned by method
    pPtr->Simple5(lArg1,&lArg2,&lArg3);
        //note that returned arguments must pass their addresses
```

Note: Even if you are ignoring an argument for a particular call, you must still NULL out that argument if it's been declared in the IDL file as a pointer ([out] or [in,out]). If there's garbage in an argument, COM will attempt to transmit whatever that garbage is pointing to.

---

3. In the server, use:

```
long lArg1,long* lArg2,long* lArg3
```

Example 31   Passing Data to a COM Object Using C++   **249**

Pass Simple Arrays

Simple arrays are those that you define using C/C++ use brackets (i.e., `stuff[3]`). Simple arrays don't work with VB but they do work with VJ++. For VB, you should instead use what's called a SafeArray as seen later in this example. For simple arrays, you can really optimize your COM communications by telling COM exactly how much of the array to transmit.

1. To pass a fixed size array in the IDL file, specify:

```
        Array1([in] long aArg[25]);
```

2. In the client, specify:

```
    long aArg[ARRAYSIZE];
    pPtr->Array1(aArg);
```

3. And in the server, use:

```
 Array1(long aArg[25]);
```

4. To pass a variable sized array in the IDL file, specify:

```
    Array2([in] long lSize, [in,size_is(lSize)] long aArg[][5]);
```

where `lSize` specifies the size of the missing dimension in `aArg[][5]`. The `aArg` argument can have only one missing dimension.

5. In the client, specify:

```
    long lSize=ARRAYSIZE;
    long vArg[ARRAYSIZE][5];
    pPtr->Array2(lSize,  (long **)vArg                           );
```

6. And in the sever, use:

```
STDMETHODIMP CWzd::Array2(long lSize, long aArg[][5])
{
    for (int i=0;i<lSize;i++)
    {
        int j=aArg[i][0];
    }
    return S_OK;
}
```

7. To specify that only a certain part of an array needs to be transmitted or received, specify in the IDL file:

```
        Array3([in] long lFirst, [in] long lLast, [in] long lSize,
[in,first_is(lFirst),last_is(lLast),size_is(lSize)] long *aArg);
```

OR

```
        Array4([in] long lFirst, [in] long lLength, [in] long lSize,
[in,first_is(lFirst),length_is(lLength),size_is(lSize)] long *aArg);
```

where lFirst is the first element to transmit, lSize is the size of the entire array and you can use either lLast to specify the last element to send or lLength to specify how many actual elements to send starting at lFirst.

Note: Although the whole array is not transmitted from client to server, the whole array is still allocated in the server.

---

8. In the client, use:

```
    long lFirst=20;
    long lLast=23;
    pPtr->Array3(lFirst, lLast, lSize, aArg);
```

OR

```
    pPtr->Array4(lFirst, lLength, lSize, aArg);
```

9. And in the server, use:

    In the IDL file for in's and out's, use:

```
STDMETHODIMP CWzd::Array3(long lFirst, long lLast, long lSize, long
    *aArg)
{
    for (int i=lFirst;i<=lLast;i++)
    {
        int j=aArg[i];
    }
    return S_OK;
}
```

Example 31   Passing Data to a COM Object Using C++   **251**

OR

```
STDMETHODIMP CWzd::Array4(long lFirst, long lLength, long lSize, long
   *aArg)
{
    for (int i=lFirst;i<(lFirst+lLength);i++)
    {
        int j=aArg[i];
    }
    return S_OK;
}
```

Pass Structure and COM Classes

Although a structure is now not that much different than a class in C++, you can't pass a regular C++ class using COM. Instead, you can only pass a pointer to a COM class.

1. To pass a structure using COM, you need to define that structure at the top of the IDL file as seen here, then use the declaration below. As for COM classes, because all COM classes are derived from the `IUnknown` class, use that argument type:

```
// define structure above any interface definitions:
typedef struct
{
    long lElement;
    long *pPointer;
    float fElement;
} MYSTRUCT;
     :    :    :
    [in] MYSTRUCT myStruct, [in] IUnknown *myClass
```

2. The structure you defined in your IDL file will be added to the `.h` file or type library along with the definitions of the methods in the COM server, so you can just start using it in your client like so:

```
MYSTRUCT myStruct;
myStruct.pPointer=new long[12];
```

```
// All embedded pointers must receive a valid
        // pointer or a NULL before it can be transmitted
    myStruct.pPointer[3]=1234;
    pPtr->Structs1(myStruct,pPtr.GetInterfacePtr());
```

> **Note:** In this particular example, the structure contained a pointer. Because COM will even transmit what an embedded pointer is pointing to, you must make sure all embedded pointers are nullified if unused too. How does COM know how much data a pointer is pointing to? By examining the memory to see how much is allocated and copying that amount.

3. And in the server, use:

```
STDMETHODIMP CWzd::Structs1(MYSTRUCT myStruct, IUnknown *myClass)
{
    IWzd *pWzd=(IWzd*)myClass;
    return S_OK;
}
```

Please see the next example for how to pass your COM pointer in a multitasking environment.

Pass Encapsulated Unions

An encapsulated union allows you to put several different data types in the same union but pick at runtime what exact type is in there so that COM can determine how many bytes to transmit. Variants operate along similar lines, however an encapsulated union can be much more diverse. You can, for instance, unionize a structure and an integer. Encapsulated unions don't work with late binding however so you're stuck with variants for VB and VJ++.

1. As with structures, you define an encapsulated union at the top of your IDL file like so:

```
// define above any interface definitions:
typedef [switch_type(long)] union //"switch_type()" makes switch a
    "long"
{
    [case(1)]
```

Example 31   Passing Data to a COM Object Using C++   **253**

```
        float fFloat[2];
    [case(25)]
        double dDouble;
    [case(27)]
        MYSTRUCT myStruct;
    [default]
                long lLong;
} MYEUNION;
    :    :    :
        [in] long lType, [in,switch_is(lType)] MYEUNION myEUnion
```

2. In the client, use:

```
    MYEUNION myEUnion;
    long lType=1;
    myEUnion.fFloat[0]=123.0f;


    pPtr->EUnions1(lType,myEUnion);
```

3. And in the server, use:

```
STDMETHODIMP CWzd::EUnions1(long lType, MYEUNION myEUnion)
{
    if (lType==1)
        float f=myEUnion.fFloat[0];

    return S_OK;
}
```

Pass Memory Pointers

When it comes to memory pointers, COM must not only transmit the pointer but also any memory the pointer was pointing to. It determines how much to send based on how much memory is allocated to the pointer. Theoretically, a memory pointer can be pointing to any kind of memory. However, if there's the potential that the server is going to deallocate the memory that the client allocated, COM provides a set of memory APIs called ::CoTaskMemAlloc() and ::CoTaskMemDealloc(). Not only do these functions keep memory from getting lost between objects, but they also create memory that's thread safe (two threads working with the same memory pointer can't write to the memory pointer at the same time).

There are three attributes you can apply to a memory pointer argument: ref, unique, and ptr:

1. "Ref" is the least overhead to COM. It just passes the memory pointer and any data it points to and the server promises not to change or deallocate the memory.

2. "Unique" is the next least overhead and the default to COM. This time the server can change the memory or even deallocate it, but the client and server promise that this is the only pointer to this particular chunk of memory so that COM doesn't have to worry about reconciling what multiple pointers do to that chunk of memory while in the server before it sends it back to the client.

3. "Ptr" is the most overhead to COM, but the most transparent to you. You don't have to worry much about anything except making sure this argument is NULL if you're not using it.

4. In the IDL file, use one of the following attribute combinations:

```
HRESULT MemPtr1([in,ref] MYSTRUCT *pStruct);
HRESULT MemPtr2([in,out,unique] MYSTRUCT **pPtr);
HRESULT MemPtr3([in,out,ptr] MYSTRUCT **pPtr1,[in,out,ptr]
MYSTRUCT **pPtr2);
HRESULT MemPtr4([in,out] MYSTRUCT **pPtr1,[in,out] MYSTRUCT
**pPtr2);
```

In this example, we're sending a structure pointer we allocated with ::CoTaskMemAlloc().

Note: You can't use LPVOID to define your memory pointer because COM needs to know exactly how big an argument you're sending.

---

5. In the client, use one of these matching argument types. In this example, we're sending a structure allocated using ::CoTaskMemAlloc():

```
MYSTRUCT *pMyStruct=
          (MYSTRUCT*)::CoTaskMemAlloc(sizeof(MYSTRUCT)
pPtr->MemPtr1(pMyStruct);
pPtr->MemPtr2(&pMyStruct);
```

Example 31   Passing Data to a COM Object Using C++   **255**

```
    MYSTRUCT *pMyStruct2=pMyStruct;
    pPtr->MemPtr3(&pMyStruct,&pMyStruct2);
    MYSTRUCT *pReturned;
    pPtr->MemPtr4(NULL, &pReturned);
```

6. And in the server, use one of these match functions:

```
STDMETHODIMP CWzd::MemPtr1(MYSTRUCT *pStruct)
{
    return S_OK;
}
STDMETHODIMP CWzd::MemPtr2(MYSTRUCT **pPtr)
{
    ::CoTaskMemFree(*pPtr);
    *pPtr=(MYSTRUCT*)::CoTaskMemAlloc(
                                        sizeof(MYSTRUCT)
        //number of bytes to allocate
                                        );

    return S_OK;
}
STDMETHODIMP CWzd::MemPtr3(MYSTRUCT **pStruct1,MYSTRUCT **pStruct2)
{
    (*pStruct1)->lElement=123;
    (*pStruct2)->lElement=456;
    return S_OK;
}
STDMETHODIMP CWzd::MemPtr4(MYSTRUCT **pStruct1,MYSTRUCT **pStruct2)
{
    return S_OK;
}
```

Pass BSTRs, SafeArrays, Variants, and Retvals

Binary strings (BSTRs), SafeArrays, variants, and return values (retvals) are the native types of Visual Basic. VB stores its text strings in BSTRs which contain Unicode characters. VB arrays are stored in SafeArrays. Untyped

VB variables are stored in variants. And when your COM object is returning a value as seen here, you need to specify it as a retval:

```
Myval = server.method()
```

Although COM would like to allow you to freely communicate between any COM object, if you don't speak the language of VB, you won't be able communicate very well with it. (For much more on communications between a VB client and a VC++ server, please refer to Example 33, page 262.)

1. In the IDL file, use the following syntax:
   - BSTRs

```
HRESULT VBVars1([in] BSTR bstr, [out] BSTR *pBSTR);
```

   - SafeArrays

```
HRESULT VBVars2([in] SAFEARRAY(BYTE) *pSA1, [out]
SAFEARRAY(BYTE) *pSA2);
```

   - Variants

```
HRESULT VBVars3([in] VARIANT VAR1, [out] VARIANT *pVAR2);
```

   - Retvals

```
"myval=server.method()"
    HRESULT VBVars4([out,retval] long *pVal);
```

Note: The "retval" attribute must be on the last argument in the method defintion.

---

2. In the client, use the following syntax:
   - BSTRs

```
// first convert string to BSTR using C++ class "_bstr_t"
  _bstr_t bstrtMyString1("MyString");
  BSTR bstrMyString2=NULL;
  pPtr->VBVars1(bstrtMyString1,&bstrMyString2);
  // convert BSTR from unicode to asciiz using W2A() macro
  USES_CONVERSION;
```

Example 31 Passing Data to a COM Object Using C++ **257**

```
char *pStr = (char*)W2A(bstrMyString2);
// or make same conversion using %S printf format
CString str;
str.Format("%S",bstrMyString2);
```

* SafeArrays

```
// create safearray and initialize with data
SAFEARRAY *saMySafeArray=
    ::SafeArrayCreateVector(
        VT_UI1,          // type (VT_UI1 == one byte unsigned
integers (bytes)
        0,               // lower index of array (can be negative)
        100);            // size
LPBYTE lpByte=NULL;
  ::SafeArrayAccessData(
        saMySafeArray,        //safearray vector from above
        (LPVOID*)&lpByte);              //pointer
  lpByte[0]=12;
lpByte[1]=34;
  ::SafeArrayUnaccessData(
        saMySafeArray);       //safearray vector from above
  pPtr->VBVars2(&saMySafeArray);
  // retrieve the data
  ::SafeArrayAccessData(
        saMySafeArray,        //safearray vector from above
            (LPVOID*)&lpByte);          //pointer
  BYTE by=lpByte[0];
  ::SafeArrayUnaccessData(
        saMySafeArray);       //safearray vector from above

  // destroy the safe array
  ::SafeArrayDestroy(saMySafeArray);
```

- Variants

```
// convert a variable to a variant using the C++ class "_variant_t"
long lLong=14;
bstrtMyString1=L"test"; //reusing a _bstr_t class
_variant_t vartMyVariant1(lLong);
_variant_t vartMyVariant2(bstrtMyString1);
// convert a safearray to a variant using V_xxx macros (other
types can be found in OLEAUTO.H)
saMySafeArray=::SafeArrayCreateVector( VT_UI1,0,100);
VARIANT varMyVariant3;
VariantInit(&varMyVariant3); //initialize to no type
V_VT(&varMyVariant3)= VT_ARRAY | VT_UI1;
V_ARRAY(&varMyVariant3)=saMySafeArray;
pPtr->VBVars3(&vartMyVariant1,&vartMyVariant2,&varMyVariant3);
// convert a variant back to a variable using "_variant_t"
lLong=long(vartMyVariant1);
bstrtMyString1=_bstr_t(vartMyVariant2);
// convert safearrays using V_xxx macros (other types can be
found in OLEAUTO.H)
if (V_VT(&varMyVariant3)==(VT_ARRAY | VT_UI1))
    saMySafeArray=V_ARRAY(&varMyVariant3);
::SafeArrayDestroy(saMySafeArray);
```

- Retvals

```
long val=pPtr->VBVars4();
// same exact method call without the IWzdPtr wrapper method
pPtr->raw_VBVars4(&val);
```

3. In the server, use these argument types:
   - BSTRs

```
BSTR bstr, BSTR* pBSTR
```

   - SafeArrays

```
SAFEARRAY** pSA
```

for both input and output — completely different from IDL syntax.

Example 31   Passing Data to a COM Object Using C++   **259**

- Variants

```
VARIANT VAR1, VARIANT* pVAR2
```

- Retvals

```
    long *pRetVal
```

or whatever the return type is.

Notes

- If the MIDL compiler balks at your selection of argument types because they aren't supported by OLE Automation, you can still use those types but you must build and register the proxy/stub DLL for your server. That is unless your COM server lives in a DLL that will be used directly by your client (no MTS/COM+, no DLLHost) and your server won't be using COM to keep it thread safe in a multitasking application. Of course, if your client is using OLE Automation (late binding), you better pick argument types it likes.

- If you forget and don't build the proxy/stub DLL and your server does need it, this is the error message you'll get, "nothing". In fact, COM will just try to muddle along without. As an example, in the case of a simple array, COM will send over the first element and then call it a day. You'll spend hours of fun playing around with your array attributes until you finally decide to build the proxy/stub DLL and register it on the client and server machines.

- The _bstr_t and _variant_t classes used above require you to include comdef.h in your source. The USES_CONVERSION macro requires atlconv.h.

- For more on this, please refer to Chapter 2.

CD Notes

- When executing the project on the accompanying CD, watch as the arguments are sent between client and server. There'll be a lot of MIDL warnings when building this project — that's because we present all IDL argument types, not just those types supported by Automation.

II

9

Example 32   Passing Interface Pointers Between Threads Using C++

Objective

You would like to safely pass a COM object pointer in a multitasking environment.

Strategy

As discussed in Chapter 3, COM can be used in some situations to provide thread-safety for your COM objects. In other words, you don't need to use critical sections, etc. to protect your data against simultaneous access by multiple threads. COM does this by using marshalling your method calls if you try to access a COM object outside of your thread. If, however, you create a COM object in your own thread, your method calls will not be marshalled. But what happens if you pass that COM object pointer to another thread? That un-marshalled pointer which won't keep your data thread safe? If you want to pass such a pointer you can use two COM API calls to add marshalling to that pointer. The API calls are: `CoMarshalInterThreadInterfaceInStream()` to pack up the pointer before the call, and `CoGetInterfaceAndReleaseStream()` to extract the pointer on the other side. Actually, COM doesn't necessarily add marshalling to the pointer — even though you bothered to type in that whole API call. COM first determines whether or not marshalling is necessary and if not, the pointer may pass through unmodified.

Steps

Pack the Pointer up for the Call

1. To pass a COM object pointer, use :`CoMarshalInterThreadInterface-InStream()` to first put it in stream format like so:

```
LPSTREAM lpStream=NULL;
hr=::CoMarshalInterThreadInterfaceInStream(
        __uuidof(IWzd),//interface ID of pointer we're passing
        pPtr.GetInterfacePtr(),//pointer to be passed
        &lpStream             //returned stream variable
```

Example 32   Passing Interface Pointers Between Threads Using C++   **261**

```
            );
    if (FAILED(hr))
    {
        _com_error err(hr);
        AfxMessageBox(err.ErrorMessage());
        return;
    }
```

2. The `lpStream` variable can now be passed freely to a function in any thread. In this example, we're using MFC's thread helper function to create a thread and send the pointer there:

```
    m_ThreadData.lpStream=lpStream;
    AfxBeginThread(
WzdThread,                      // static thread process declared
                                // as (UINT WzdThread( LPVOID pParam );)
    &m_ThreadData);             // data to send to thread
```

II

Extract the Pointer

1. In the called function, you can extract the COM object pointer with:

```
    LPUNKNOWN p;
    HRESULT hr=::CoGetInterfaceAndReleaseStream(
        pData->lpStream,            //stream
        __uuidof(IWzd),             //identifier of the interface
        (LPVOID*)&p);
    if (FAILED(hr))
    {
        _com_error err(hr);
        AfxMessageBox(err.ErrorMessage());
        return 0;
    }
```

9

2. The pointer comes out as an `LPUNKNOWN`, so stick it in a smart pointer with:

```
    IWzdPtr pPtr(p);
```

3. As mentioned above, `pPtr` may or may not now be a marshalled pointer.

Notes

- Please see Chapter 3 for much more on how COM can be used to automatically make your COM objects thread safe.

CD Notes

- When executing the project on the accompanying CD, you will notice that the pointer is unmarshalled in the main application, but marshalled in the thread.

Example 33   Passing Data Between C++ and Visual Basic

Objective

You would like to exchange data between your C++ COM object and your Visual Basic client.

Strategy

In this example, we will simply be reviewing the subset of COM argument types you can safely use with a Visual Basic client, from IDL syntax in the COM server to the VB argument types they correspond with.

Steps

Pass Simple Data Types

To pass bytes, integers, longs, singles (floats), and doubles:

1. In VB, use:

```
Dim cArg As Byte
Dim sArg As Integer
Dim lArg As Long
Dim fArg As Single
Dim dArg As Double
cls.method(cArg, sArg, lArg, fArg, dArg)
```

Example 33 Passing Data Between C++ and Visual Basic **263**

2. In the IDL file in the COM server, use:

```
    [in] unsigned char cArg, [in] short sArg, [in] long lArg,
[in] float fArg, [in] double dArg
    [out] unsigned char *pcArg, [out] short *psArg, [out] long *plArg,
[out] float *pfArg, [out] double *pdArg
```

3. In the .cpp and .h C++ files of the COM server, use:

```
unsigned char cArg, short sArg, long lArg, float fArg,
double dArg
unsigned char *pcArg, short *psArg, long *plArg,
float *pfArg, double *pdArg
```

Pass Strings

To pass a Visual Basic string, which to C++ is a binary string allocated with the Win32 API calls SysAllocString() and SysFreeString():

1. In VB, use:

```
Dim sArg As String
cls.method(sArg)
```

2. In the IDL file of the server for in's and out's, use:

```
[in] BSTR bstr
[out] BSTR *pBSTR
```

3. In the .cpp and .h C++ files of the server, use:

```
BSTR bstr
BSTR *pBSTR
```

Pass Arrays

To pass a VB array, which in C++ is a SafeArray, use:

1. In VB use:

```
Dim aArg(1 To 25) As Byte
cls.method(aArg)
```

2. In the IDL file of the C++ server for in's and out's, use:

```
[in] SAFEARRAY(BYTE) sa
[out] SAFEARRAY(BYTE) *pSA
```

3. In the `.cpp` and `.h` C++ files, use:

```
SAFEARRAY* sa
SAFEARRAY** pSA
```

Note: You will find examples of allocating and deallocating a SafeArray in Example 31 (page 246).

Pass Variants

To pass variants, which is a VB variable that you don't type or type as a Variant:

1. In VB, use:

```
Dim vArg As Variant    'or any untyped argument
cls.method(vArg)
```

2. In the IDL file for in's and out's, use:

```
[in] VARIANT var
[out] VARIANT* pVAR
```

3. In the `.cpp` and `.h` C++ files, use:

```
VARIANT var
VARIANT* pVAR
```

Return Values

To get a returned variable from the method call (you'll see below):

1. In VB, use:

```
Dim lArg as Long
lArg = cls.method()
```

2. In the IDL file, use (at the end of the line of arguments!):

```
[out,retval] long *plArg
```

3. In the `.cpp` and `.h` files, use:

```
long *plArg
```

Notes

- You don't need a proxy/stub DLL to talk to VB or VJ++ because the COM DLL that implements the COM API knows how to send a standard set of argument types and it finds out what arguments are in your method calls from your COM server's type library.
- Don't worry about deallocating binary strings or SafeArrays passed to your COM object from VB. In most cases, VB will clean these up itself and will, in fact, get upset if you do it first.
- For more notes, please refer to the end of Example 31 (page 246).

Example 34   Passing Data Collections Between Visual C++ & Visual Basic

Objective

You would like to pass a data collection from your C++ COM server to a VB client.

Strategy

A data collection in this case is the VB variable type that allows you to do this in VB:

```
For Each MyObject In MyCollection
    ' use MyObject, which could be a safearray, BSTR, etc.
Next
```

Where "MyCollection" is the data collection. In fact, the data collection that we're passing from our COM server is itself a COM server that we need to implement with a set of methods if we expect VB to be able to use it. Inside of this data collection server is an array or list of data stored any way we see fit.

In lieu of reviewing all of the gory details of what goes on inside of this data collection server, we'll briefly review what goes on and you can grab the code from the CD and modify it for your own project.

Steps

Write the Data Collection Server

A data collection server will generally have five methods to which you can add other custom methods. The Item() method return the data object at an index. The Add() method adds a new object to the collection. The Init() method can initialize the maximum size of the collection — or you can just make Add() smart enough to keep appending the list. The Count() method returns the current number of elements in the collection. And the _NewEnum() method is used by VB in the For...Next loop we saw above.

The IDL file for this server looks like this:

```
HRESULT Item([in] VARIANT Index, [out, retval] LPVARIANT pItem);
HRESULT Add([in]SAFEARRAY(unsigned char)* NewObj);
HRESULT Count([out, retval] long *pVal);
HRESULT Init([in]long numObjs);
HRESULT _NewEnum( [out, retval] LPUNKNOWN* pVal);
```

1. Copy the Data Collection Server from the CD and modify it for your application. A sample implementation can be found in the listing at the end of this example.

Use the Data Collection Server in Your Own Server

1. Specify the collection argument in the IDL file of your real COM server (the one that's going to be passing the data collection to VB):

```
[out] LPUNKNOWN *iCollection
```

2. Create the collection server inside of your real server and initialize it to the number of objects it will contain:

```
IWzdCollectionPtr pPtr;
pPtr.CreateInstance(__uuidof(WzdCollection));
pPtr->Init(
            3                   // number of objects to add
            );
```

3. Add those objects to the collection. In this case, we're adding SafeArrays:

```
for (int i=0;i<3;i++)
{
    // create object (safearray, BSTR, etc.)
    LPBYTE lpByte;
    SAFEARRAY *pSA=::SafeArrayCreateVector(VT_UI1,0,10);
    ::SafeArrayAccessData(pSA,(LPVOID*)&lpByte);
    lpByte[0]=(BYTE)i;
    ::SafeArrayUnaccessData(pSA);
    pPtr->Add(&pSA);
}
```

4. Before returning the collection, don't forget to detach it from the smart pointer — otherwise the smart pointer will release it before it even gets back to your VB client:

```
*pCollection=pPtr.Detach();
```

Use the Data Collection Server in Your VB Application

1. At the same time you're adding the real COM server to your VB app, you will also need to add the Data Collection COM server using the "Project/Reference" menu command.

2. Create the COM server as usual:

```
Dim IWzdSrv As Object
Set IWzdSrv = CreateObject("Server.Wzd")
```

3. Define the collection server just before you call the real server:

```
Dim MyCollection As WzdCollection
```

4. And also define a variable that will contain each object as it comes out of the collection:

```
Dim MyObject
```

5. Use your real server to get the collection server (and its contents):

```
rVal = IWzdSrv.GetCollection(MyCollection)
```

II

9

6. To access individual items in the collection, you can use the `Item()` method:

```
MyObject = MyCollection.Item(0)
```

7. But to use the collection as it was intended, use:

```
For Each MyObject In MyCollection
    ' use MyObject, which could be a safearray, BSTR, etc.
Next
```

Listings — Collection Server Class

```cpp
// WzdColl.cpp : Implementation of CWzdCollection
#include "stdafx.h"
#include "CollSrv.h"
#include "WzdColl.h"


//////////////////////////////////////////////////////////////////////
    ////////
// CWzdCollection
typedef
    CComObject<CComEnum<IEnumVARIANT,&IID_IEnumVARIANT,VARIANT,
    _Copy<VARIANT> > > EnumVar;

// initilize number of safe arrays
STDMETHODIMP CWzdCollection::Init(long numObjs)
{
    m_inx=0;
    m_numObjs=numObjs;
    m_var=new VARIANT[numObjs];
    for (int i=0; i<numObjs; ++i)
        VariantInit(&m_var[i]);
    return S_OK;
}

// add another safe array
STDMETHODIMP CWzdCollection::Add(SAFEARRAY **ppNewObj)
{
```

```
        if (m_inx<m_numObjs)
        {
            m_var[m_inx].vt = VT_ARRAY | VT_UI1;
            m_var[m_inx].parray = *ppNewObj;
            m_inx++;
            return S_OK;
        }
        return E_FAIL;
}


// return count
STDMETHODIMP CWzdCollection::get_Count(long *pVal)
{
    *pVal=m_numObjs;
    return S_OK;
}


// return enum
STDMETHODIMP CWzdCollection::get__NewEnum(LPUNKNOWN *pEnumObjs)
{
    EnumVar *pVar=new EnumVar;
    pVar->Init(&m_var[0],&m_var[m_numObjs],NULL,AtlFlagCopy);
    pVar->QueryInterface(IID_IUnknown,(void**)pEnumObjs);
    return S_OK;
}


// return one item
STDMETHODIMP CWzdCollection::Item(VARIANT Index, LPVARIANT pItem)
{
    pItem->vt- VT_ARRAY | VT_UI1;
    pItem->parray = m_var[Index.lVal].parray;
    return S_OK;
}


// cleanup on destroy
void CWzdCollection::FinalRelease()
{
```

II

9

```
if (m_var)
{
    for (int i=0; i<m_inx; ++i)
    {
        SafeArrayDestroyData(m_var[i].parray);
        m_var[i].vt = VT_EMPTY;
    }
    delete []m_var;
}
}
```

Example 35   Passing Data Between C++ and Visual J++

Objective

You would like to exchange data between your C++ COM object and your Visual J++ client.

Strategy

In this example, we will simply be reviewing the subset of COM argument types you can safely use with a Visual J++ client, from IDL syntax in the COM server to the VJ++ argument types they correspond with.

Steps

Pass Simple Data Types

To pass chars, shorts, integers, floats, and doubles into C++:

1. In VJ++, use these argument types:

```
char byArg='a';
short sArg=1234;
int lArg=1234567890;
float fArg=1.234f;
double dArg=1.23456789;
wzd.method(byArg,sArg,lArg,fArg, dArg);
```

Example 35   Passing Data Between C++ and Visual J++   **271**

2. In the C++ server's IDL file, use:

```
    [in]  char byArg, [in] short sArg, [in] long lArg,
[in] float fArg, [in] double dArg
```

3. And in the C++ server, use:

```
BYTE byArg, short sArg, long lArg, float fArg, double dArg
```

Pass and Receive Data with [in] and [out]

In the last section, we were just passing data from our VJ++ client to a C++
COM server. In this section, we will look at the IDL argument attributes `in`
and `out` and what VJ++ syntax is necessary to receive a variable back from
C++. Although we only review the `int` data type here, this syntax applies to
all types:

1. In VJ++, use:

```
        int lArg1=1234;                      //[in]
        int lArg2[]=new int[1];              //[out]
        int lArg3[]=new int[1];              //[in,out]
        lArg3[0]=5678;
        wzd.method(lArg1, lArg2, lArg3);
```

2. In the C++ server's IDL file, use the matching argument from here:

```
    [in] long lArg1,[out] long *lArg2,[in,out] long *lArg3
```

3. And in the server's `.cpp` and `.h` files, use:

```
    long lArg1,long* lArg2,long* lArg3
```

Pass Simple Arrays

Although simple arrays (i.e., `data[23]`) don't work with Visual Basic, they
are supported by the COM DLL for late binding and can be used by VJ++
clients:

1. In VJ++, use:

```
        int lFirst=20;
          int lLast=23;
          int lSize=25;
          int    aArg[]=new int[lSize];
          wzd.method(lFirst, lLast, lSize, aArg);
```

OR

```
        int lLength=3;
        wzd.method(lFirst, lLength, lSize, aArg);
```

2. In the server's IDL file, use:

```
    [in] long lFirst, [in] long lLast, [in] long lSize,
[in,first_is(lFirst),last_is(lLast),size_is(lSize)] long *aArg);
```

OR

```
    [in] long lFirst, [in] long lLength, [in] long lSize,
    [in,first_is(lFirst),length_is(lLength),size_is(lSize)] long
    *aArg);
```

3. And in the server's .cpp and .h files, use:

```
long lFirst, long lLast, long lSize, long *aArg
```

OR

```
long lFirst, long lLength, long lSize, long *aArg
```

where the elements from lFirst to lLast are sent and the array, aArg, is lSize elements big. Or in the second case, lFirst is the first element sent and lLength is the number of elements after that are sent.

Pass BSTRs

A BSTR argument type is the binary string type that's native to Visual Basic. However it's also implemented in VJ++ as a String class:

1. In VJ++, use:

```
        String bstr="Test";
        String[] pBSTR=new String[1];
        wzd.method(bstr, pBSTR);
```

2. In the server's IDL file, use:

```
        [in] BSTR bstr, [out] BSTR *pBSTR
```

3. And in the server's .cpp and .h files, use:

```
        BSTR bstr, BSTR* pBSTR
```

Example 35   Passing Data Between C++ and Visual J++   **273**

Pass SafeArrays

SafeArrays are VB's native format for arrays. In order to pass a SafeArray to VJ++, we need to include the Microsoft COM package for a special SafeArray class:

1. In VJ++, use:

```
import com.ms.com.SafeArray;
    :     :    :
        SafeArray psa = new SafeArray(Variant.VariantByte,100);
        psa.setByte(0,(byte)12);
        wzd.method(psa);
```

2. In the server's IDL file, use:

```
        [in,out] SAFEARRAY(BYTE) *pSA
```

3. And in the server's `.cpp` and `.h` files, use:

```
        SAFEARRAY** pSA
```

Pass Variants

The Variant class is also included with the Microsoft Java package, which allows VJ++ to accept a variant argument from a COM server:

1. In VJ++, use:

```
import com.ms.com.Variant;
        Variant pVAR1=new Variant(1234);
        Variant pVAR2=new Variant(bstr);
        Variant pVAR3=new Variant(psa);
        wzd.method(pVAR1, pVAR2, pVAR3);
```

2. In the server's IDL file, use:

```
[in,out] VARIANT *pVAR1,[in,out] VARIANT *pVAR2, [in,out] VARIANT
    *pVAR3
```

3. And in the server's `.cpp` and `.h` files, use:

```
VARIANT* pVAR1,VARIANT* pVAR2,VARIANT* pVAR3
```

II

9

Get a Return Value

Because COM reserves the return value for its own error reporting, we need to use the "retval" argument attribute in order to have our own value returned instead. Actually, the compiler simply moves the argument in question out of the argument list to the front of the method:

1. In VJ++, use:

```
int retval=wzd.method();
```

2. In the IDL file, use:

```
[out,retval] long *pVal
```

which must be at the end of the argument list.
And in the `.cpp` and `.h` C++ files, use:

```
..., long *pVal);
```

or whatever argument type you are using.

Notes

- By comparing this example with the Visual Basic example earlier in this chapter, you can also figure out how to pass data between a VB client and a VJ++ COM object. Some pundits suggest this as the ideal configuration — a simple to program front end using VB using robust VJ++ COM objects.

CD Notes

- Create both the VJ++ and VC++ projects. Tell the VC++ project, which is an ATL COM DLL, to use the VJ++ exe to run it when debugging. You can configure this in the "Debug" tab of the project settings in the "Executable for debug" edit box. Then, set a breakpoint in each method of the ATL COM object and watch as the VJ++ application sends it data.

Chapter 10

COM+ Examples

COM+ represents the frontier of COM development. Contrary to what its name implies, COM+ isn't an overall enhancement of COM but rather an enhancement to a certain area of COM development. On the other hand, some would insist it makes the rest of COM irrelevant, but even its strongest supporters will be out there writing COM NT Services or regular COM objects when needed.

What area does COM+ enhance? For the complete story, please turn back to Chapter 4, but in general, COM+ is targeted towards database applications that require lots of scalability (read: can support ten thousand clients as easily as one). In fact, some would consider COM+, which is only available on Windows 2000, to simply be the next version of the Microsoft Transaction Server (MTS), a service available from the NT Options pack that runs on NT. Because we're in that delicate transition period between NT and Windows 2000, I have included examples for both in this chapter.

One of the more striking characteristics you will find in the following examples is that they contain little or no code you need to write yourself. Instead you are simply activating functionality that COM+ has available to your object. When the options outnumber the lines of code, you're talking about "attribute" programming. Soon 99% of every book on computers will be on how to setup your options and your code will be size lines of Basic.

The examples in this chapter include:

Example 36   Writing an MTS or COM+ Server Using ATL   where we write an ATL C++ COM server that fully utilizes an MTS or COM+ environment.

Example 37   Registering a Server with MTS   where we review the steps necessary to register your COM server with MTS.

Example 38   Registering a Server with COM+   where we do the same for COM+.

Example 39   Using the COM+ Event Server   where we outline how to use COM+'s new event service to send an event to multiple servers at once.

Example 40   Writing and Using a COM+ Queued COM Server
where we itemize what's necessary to write a COM server that will queue up its input until the machine that it's on is available.

Example 36   Writing an MTS or COM+ Server Using ATL

Objective

You would like to write a COM object that can take full advantage of the MTS or COM+ DLL surrogate environment.

Strategy

In order for your COM DLL server to take full advantage of the MTS or COM+ surrogate environment, you need to add support for two objects to your server: `IObjectContext` and `IObjectControl`. By using the methods of `IObjectContext`, you are able to take advantage of MTS/COM+'s transaction support that allows you to commit or rollback transactions to multiple databases over multiple objects. And by implementing `IObjectControl` in your own server, you're able to take advantage of COM+'s object pooling functionality.

Example 36   Writing an MTS or COM+ Server Using ATL   **277**

We will be using ATL's wizards to automatically add this support to our server. We will also look at how to write an MTS/COM+ object so that it works optimally in that environment. Think "elaborate stored procedure".

Steps

Creating the Project and Class

1. Click on the Studio's "File" and "New" menu commands to open the New dialog. Then, pick "ATL COM AppWizard" and enter a project name to open the ATL AppWizard.

2. You can only create a dll project. You must also pick "Support MTS". If you plan on having any non-standard argument types in your methods (e.g., structures, or see list in Chapter 2), then also select "Allow merging of proxy/stub code" to make your life a lot easier. (You won't have to register your MTS/COM+ dll server *and* your proxy/stub dll on each machine.) You can also add MFC support, but MTS/COM+ objects don't tend to have any user interface involvement and the other MFC classes like CList can substituted for their smaller Standard Template Library (STL) equivalents.

3. To add a class to this project, click on the Studio's "Insert" and "New ATL Object" menu commands to open the ATL Object Wizard. From the list of objects, pick the "MTS Transaction Server". (This is with v6.0 of the Studio. Version 7.0 will mention COM+.)

4. You can now name your MTS/COM+ class (remember it can't be the same name as the project). On the "MTS" tab, you can add support to your server to be usable by early- and late-binding clients. Just leave it as "Dual". Another option is to support "IObjectControl". This option adds an implementation to your DLL server for the IObjectControl class which is what COM+ uses to determine and control your server in a pooled object environment. This option is meaningless in MTS because MTS doesn't support object pooling.

Writing the Class Methods

1. Once your MTS/COM+ project and class are created, you can add methods to your class just like any other ATL class by right-clicking on the interface in the ClassView and selecting "Method". For more details, please see Example 18 (page 192).

2. When dividing your functionality up into methods, try to put all related activity into individual methods. Methods should be written like really smart stored procedures that get in, do a couple of database transactions, and leave. Don't save anything in a member variable for next time, because your object will be destroyed immediately upon return to the client anyway. And at the end call SetAbort() or SetComplete() to respectively rollback or commit your changes to the database. Your method should look something like this:

```
STDMETHODIMP CWzd::Method1(long lArg)
{
    HRESULT hr=S_OK;
    try
    {
        // use ADO to access database


        // create other MTS objects

    }
    catch(_com_error& e)
    {
        hr=e.Error();
    }
    catch (...)
    {
        hr=E_FAIL;
    }


    // if context object exists (we're not in debug environment)
    if (m_spObjectContext)
    {
        if(FAILED(hr))
            m_spObjectContext->SetAbort();
        else
```

Example 36   Writing an MTS or COM+ Server Using ATL   **279**

```
            m_spObjectContext->SetComplete();
    }
    return hr;
}
```

3. With COM+, you also have the additional option to call `m_spObjectContext->DisableCommit()`. By calling this method, you prevent your object from being destroyed immediately upon return from a method call. You can then use member variables in your object to save data between method calls. To finally allow your object to commit its changes, you would then call `m_spObjectContext->EnableCommit()`.

4. If you are creating a COM+ server that will be poolable, you need to also implement the `Activate()` and `Deactivate()` methods of the `IObject-Control` class you had the wizard add above. `Activate()` and `Deactivate()` should be implemented just like a constructor and destructor, initializing and destroying the member variables of your class.

5. Please refer to the end of this example for a full list of MTS/COM+ class.

Notes

- You should also add some type of error reporting capability to your class along the lines of Example 46 (page 333). When you have objects flying around in a release environment, you could have sporadic bugs happening that could never be traced otherwise.

- Notice in the listing above that we first check if `m_spObjectContext` is null before using it. This member variable of your class is assigned to your object's `IObjectContext` (`IObjectContext` actually aggregates your object). But outside of MTS or COM+, such as when debugging your class, there won't be any `IObjectContext` and your class will blow up.

- Don't worry about adding critical sections to your method calls for thread-safety if you're using ADO or COM+. If you're using ADO to access your database(s), ADO will synchronize your updates. And COM+ pretty much forces you to use their synchronization support which is equivalent to putting your entire transaction into one big critical section.

- To use this object, first register it with the correct environment for your operating system (MTS for Windows NT or COM+ for Windows 2000). Then, create and use the object from any client the same way you would use any COM object. Please see Example 2 (page 110) for how to create the object using smart pointers.

Listings — MTS/COM+ Class

```cpp
// Wzd.cpp : Implementation of CWzd
#include "stdafx.h"
#include "Server.h"
#include "Wzd.h"
#include <comdef.h>


///////////////////////////////////////////////////////////////////////
////////
// CWzd

HRESULT CWzd::Activate()
{
    HRESULT hr = GetObjectContext(&m_spObjectContext);
    if (SUCCEEDED(hr))
        return S_OK;
    return hr;
}

BOOL CWzd::CanBePooled()
{
    return TRUE;
}

void CWzd::Deactivate()
{
    m_spObjectContext.Release();
}


STDMETHODIMP CWzd::Method1(long lArg)
{
    HRESULT hr=S_OK;
```

Example 36   Writing an MTS or COM+ Server Using ATL   **281**

```
try
{
    // use ADO to access database


    // create other MTS objects




}
catch(_com_error& e)
{
    hr=e.Error();
}
catch (...)
{
    hr=E_FAIL;
}

// if context object exists (we're not in debug environment)
if (m_spObjectContext)
{
    if(FAILED(hr))
        m_spObjectContext->SetAbort();
    else
        m_spObjectContext->SetComplete();
}
return hr;
}
```

II

10

Example 37   Registering a Server with MTS

Objective

You would like to register your DLL server with MTS.

Strategy

We will be using the Microsoft Management Console to register the COM DLL server we created above with MTS. MTS comes with the Windows NT Option pack for Windows NT. If you will be using your COM server with COM+, please refer to the next example. For Windows 9x, forget about it.

Steps

Open the Microsoft Management Console

1. The Windows NT Option pack installation puts a new menu item in your system's start menu. Underneath this item you will find the "Microsoft Transaction Explorer" which opens the "Microsoft Management Console" or MMC. The reason they're not called the same thing is that the MMC was intended to support several management applications that "snap-in" to the MMC.

Register the COM DLL Server

To register a COM DLL server with MTS you must first group them together by functionality on paper. You then create an MTS "package" for each of these groups. There are two types of packages: Server and Library. A Server package gets its own process space when running while all of the objects of a Library package will run in any Server package's process space. If you have an object that is required by more than one package, you must put it in a Library package.

> **Note:** Don't get too attached to the name "package". It changes to "application" in COM+.

1. To create a package, locate the current list of packages registered on your system under the "Computers" branch of "Microsoft Transaction

Example 37   Registering a Server with MTS   **283**

Server". Right-click on "Packages Installed" underneath and select "New". You are then led through a series of intuitive pages that help you to create the package. When selecting the "Package Identity", pick "Interactive user" only if someone is always guaranteed to be logged into the server. But don't sweat this question too much now because you can change this later.

Note: The MMC has a user-hostile interface, perhaps due to its snap-in nature. You must fully select a branch with the left cursor before you can right-click on it to get the correct popup menu.

---

2. Once the package has been added, locate it in the list of packages and right-click on it. Then, from the popup menu, select "Properties". From the "Security" tab of this property sheet, you can change the Package Identity you set above. From the "Activation" tab, you can pick what type of package this is: Server or Library. Please see Figure 10.1 for a review of these steps.

3. Next we will add our COM DLL server to the package we just created. Locate the "Component" branch under your new package and right-click on it. Select "New" to open the Component Wizard, which will lead you through the intuitive process of adding your COM DLL. Along the way, you are given the choice of adding a new component or one that is already registered on your system. If you used ATL to create your object on a machine, it will have already been registered using regsvr32. However, in most cases, you should just locate the actual DLL filename on the system so that there's no mistake as to which one you install.

II

10

Figure 10.1 Creating an MTS Package

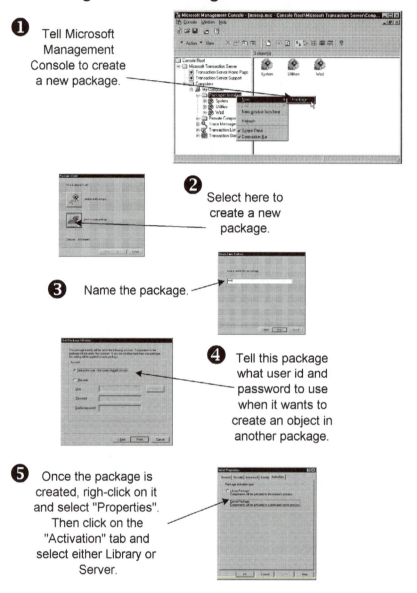

1 Tell Microsoft Management Console to create a new package.

2 Select here to create a new package.

3 Name the package.

4 Tell this package what user id and password to use when it wants to create an object in another package.

5 Once the package is created, right-click on it and select "Properties". Then click on the "Activation" tab and select either Library or Server.

4. After you've added your COM DLL to a package, you must still configure how you want MTS to support its transactions. Locate your new COM DLL under the package's components and right-click on it. Select

Example 37 Registering a Server with MTS **285**

"Properties" to edit your server's properties — one of which the type of transaction support you want from MTS where:

Transaction Option	Resulting Behavior
Requires a Transaction	When MTS creates this object, it either starts a new transaction if one hasn't been started yet, or uses the transaction that its client object started.
Requires a New Transaction	When this object is created, MTS always starts a new transaction. When this object's methods return, whatever they did to the database is committed or rollbacked right there even if its caller isn't the original client.
Supports Transactions	When this is selected, the object will pass a transaction through to any objects it creates, but it will not be involved with committing or rolling back the transaction.
Does not support Transactions	In this case, not only is the object not participating in the transaction, but it won't even pass a transaction through to any objects it creates.

5. For a review of adding a COM DLL to an MTS package, please see Figure 10.2.

Notes

- Once a DLL has been registered with MTS, you can find out whether or not its running using the Microsoft Management Console. Open the tree view on the left so that the DLL is visible as a ball in the list control on the right. When the object is running, the ball will spin. By the way, that's an "X" on the ball, a remnant of the ActiveX naming convention.

Figure 10.2   Registering Your COM DLL with MTS

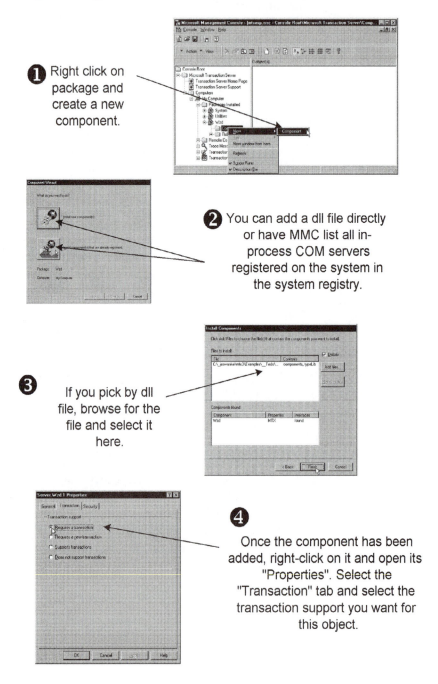

1 Right click on package and create a new component.

2 You can add a dll file directly or have MMC list all in-process COM servers registered on the system in the system registry.

3 If you pick by dll file, browse for the file and select it here.

4 Once the component has been added, right-click on it and open its "Properties". Select the "Transaction" tab and select the transaction support you want for this object.

Example 38   Registering a Server with COM+   **287**

Example 38   Registering a Server with COM+

Objective

You would like to register your DLL server with COM+.

Strategy

We will be using COM+'s Component Services snap-in to register the COM DLL server we created above. COM+ only exists on Windows 2000. For Windows NT, you will need to use the Microsoft Transaction Server (MTS) discussed in the last example. For Windows 9x, you will need to write your own DLL surrogate.

Steps

Open the Component Services snap-in

1. Unlike MTS, COM+ is part of the operating system. And its administrator application is now located in your system's control panel settings under "Component Services".

Register the Server with COM+

To register a COM DLL server with COM+, you must first group them together by functionality on paper. You then create a COM+ "application" for each of these groups. There are two types of applications: Server and Library. A Server application gets its own process space when running while all of the objects of a Library application will run in any Server application's process space. If you have an object that is required by more than one application, you must put it in a Library application.

1. To create a package, locate the current list of packages registered on your system under the "Computers" branch of "Component Services". Right-click on "COM+ Applications" underneath and select "New". You are then led through a series of intuitive pages that help you to create the application. When selecting the "Application Identity", pick "Interactive user" only if someone is always guaranteed to be logged into the server. But don't sweat this question too much now because you can modify your application's properties later.

2. If you're at all familiar with MTS registration, COM+ isn't much different. The biggest difference is that you can specify whether an application is a Server or a Library in the creation process rather than through its properties later.

3. To add your COM DLL to an application, locate the "Components" branch under your new application and right-click on it. (Note: You must fully select a branch with the left mouse button before you can right-click on it to get the correct popup menu.) Then, click on "New Component" to browse your machine for the DLL filename. See Figure 10.3.

4. If you're familiar with MTS, you'll notice a new button that allows you to register an "Event Class". Please see Example 39 (page 294) for more on this.

5. For an overview of adding a DLL server to COM+, please refer to Figure 10.4.

6. Once added to an application, you can endow your object with optional functionality from COM+ by editing its properties. To edit a dll's properties, locate it in the Component Services' tree view and right-click on it. Then, select "Properties".

7. You have the same transaction support as MTS (see the last example), plus the ability to set a timeout so that your server application doesn't hang the system. You can also enable your object to be pooled, provide it with arguments when it's created by COM+, and tell COM+ to create it when the client asks for it to be created.

8. You also have the ability to ask COM+ to make sure your object is thread safe under the "Concurrency" tab. Because there's very little overhead for this feature and it has the effect of placing your object and any object it calls into one big critical section, you should always leave it enabled.

9. For a review of the steps to configure your COM+ object, please see Figure 10.5.

Example 38   Registering a Server with COM+   **289**

Figure 10.3   Creating a New Application in COM+

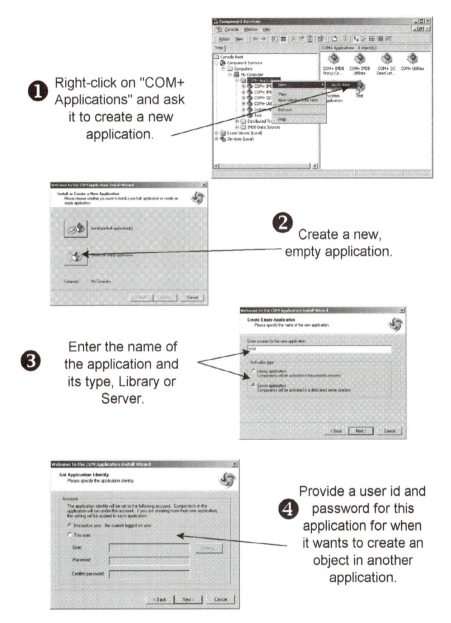

1 Right-click on "COM+ Applications" and ask it to create a new application.

2 Create a new, empty application.

3 Enter the name of the application and its type, Library or Server.

4 Provide a user id and password for this application for when it wants to create an object in another application.

II

10

Figure 10.4   Registering a COM DLL Server with COM+

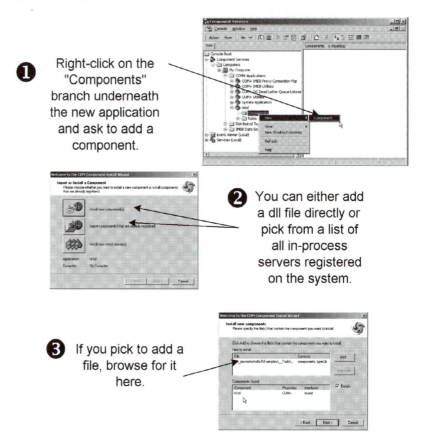

❶ Right-click on the "Components" branch underneath the new application and ask to add a component.

❷ You can either add a dll file directly or pick from a list of all in-process servers registered on the system.

❸ If you pick to add a file, browse for it here.

Adding Method Level Security with COM+

Before COM+, you could only implement security at the object level for any COM object. In other words, you could only allow or deny access to a user for all the methods of an object. With COM+, however, you are now able to allow users to access certain methods of an object while denying them access to other more dangerous methods. And rather than specify individual users for each method call, which could get pretty tedious, individual user ids are grouped into what are called "roles", such that everyone with a particular role will be allowed access to a particular method. Therefore, you start by creating these roles for your object and assigning user ids to them.

Example 38   Registering a Server with COM+   **291**

Figure 10.5   Configuring a COM DLL in COM+

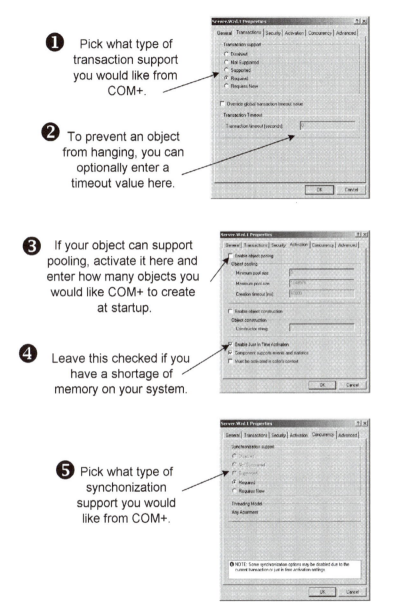

1 Pick what type of transaction support you would like from COM+.

2 To prevent an object from hanging, you can optionally enter a timeout value here.

3 If your object can support pooling, activate it here and enter how many objects you would like COM+ to create at startup.

4 Leave this checked if you have a shortage of memory on your system.

5 Pick what type of synchonization support you would like from COM+.

1. Locate the "Roles" branch underneath your application in the Component Services tree view and right-click on it to create a new role. Give this role a descriptive name — in a hospital environment, you might create a

doctor role and a nurse role, such that a doctor who logs in will be able to access more functionality than a nurse.

2. Once a role is created, locate the "Users" branch underneath it and right-click on it to start adding users to it. The users you will be adding come straight out of your system's User Manager.

3. After a role has been added to an application, you can go back to each server and even each method of a server and use this role to protect that server or method. To protect a method, locate it from under a server component and right-click on it. Then, select "Properties" and click on the "Security" tab. You will find your roles listed there. Check mark all roles that are authorized to call that method.

4. For a review of making a method secure, please see Figure 10.6.

Notes

- Actually, the security we implemented above is the last line of defense. In a well-designed system, an unauthorized user wouldn't even get to this point because the user interface would forbid it (e.g., a login system connected to an alarm system). COM+'s protection is intended to prevent unauthorized access to a more savvy user.

- With COM+ comes the new threading model: "Neutral". For more on this new model, please refer to Chapter 4. However, Neutral is not in the vocabulary of v6.0 of the Studio. Until v7.0 of the Studio makes it possible otherwise, you will need to manually edit your project's .rgs file and replace "Both" with "Neutral". Then, rebuild and register.

Example 38   Registering a Server with COM+   **293**

Figure 10.6   Configuring Security for a COM DLL Server in COM+

1 In another branch under the application called "Roles" right-click and ask to create a new role.

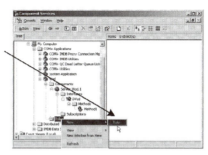

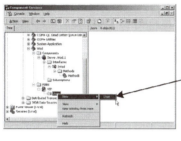

2 Then under that new role, right-click on "Users" and ask to add a new user id to this role.

II

3 Pick a user id from your Windows 2000's User Manager.

4 Then go back to the component and select a method underneath you want to protect.

10

5 And enable the role you just created.

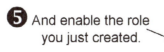

Example 39   Using the COM+ Event Server

Objective

You would like to create and call multiple COM+ objects when an event happens.

Strategy

Although this functionality is related to connection points and sinks as laid out in Chapter 2, the event server support in COM+ is more like an automated way to create and call several other COM+ objects when some event occurs. We start by creating a dummy event class which has the interface and methods that will be called but nothing else. We then create some real COM classes that pattern their interface and methods after this event class, but this time, actually implement some functionality. We tell COM+ about the event class and about the other classes that are based on it. And whenever someone creates and calls the event class, COM+ takes over and creates and calls the rest.

In COM+ terminology, the classes that pattern themselves after the event class are called "subscribers" and the entity that creates and calls the event class is called the "publisher".

Steps

Create and Register the Event Class

1. Use ATL Wizards to create an MTS/COM+ class as laid out in the first example in this chapter. Add methods to this class that you would like to be able to call when some event happens.

2. Register this class with COM+, but this time we use the new "Event Class" button. You will then be prompted to browse for the event class's DLL.

3. After the class has been registered, open its properties by right-clicking on it. Under the "Advanced" tab of the property sheet, you are given the option to allow all subscribers to be informed simultaneously or sequentially waiting for each method call to complete before the next one is called. You are also given the chance to make this event class available to other objects in the same application. For this, you will also need to specify an id to use by the subscribers.

Example 39   Using the COM+ Event Server   **295**

4. For a review of these steps, please see Figure 10.7.

Figure 10.7   Registering the "Dummy" Event Class

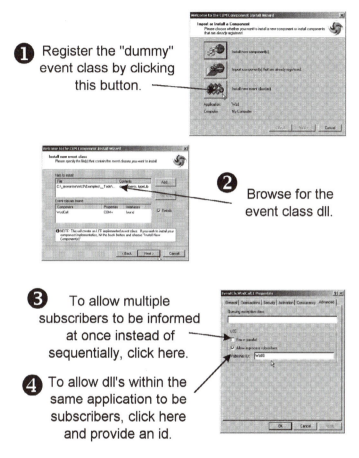

❶ Register the "dummy" event class by clicking this button.

❷ Browse for the event class dll.

❸ To allow multiple subscribers to be informed at once instead of sequentially, click here.

❹ To allow dll's within the same application to be subscribers, click here and provide an id.

Create and Register the Subscriber Class

1. Create the subscriber classes using ATL as usual. Add an interface and methods to this class that exactly duplicate those in the event class. The actual Interface ID doesn't have to be the same, but the argument lists must.

2. Register the server with COM+ as usual.

3. Locate the "Subscription" branch under this new server, right-click it, and ask to create a new subscription. Tell COM+ you want to use all interfaces for this event, select the ProgID of the event class, and give the subscription a name.

4. For a review of these steps, please refer to Figure 10.8.

Notes

- Anyone from a VB client to another COM+ object can now create and call the event class's methods and COM+ will do the rest.

Example 40   Writing and Using a COM+ Queued COM Server

Objective

You would like to write a COM server that can queue up its requests until the system it's located on is available.

Strategy

All along we've assumed that when a client wants to talk to a COM server, the machine it's sitting on is available. But what about the case of a laptop computer that might go days without being connected to any network that would have a COM server available. You could write your own queuing facility to store up data and then use COM to send it in when connected. However, objects running in the COM+ environment can use COM+ to store up its data instead.

Essentially, the client in COM+, rather than creating the COM server directly, creates a special COM+ object instead called QC.Recorder. The client then talks to this object as if it were the real thing until it's done. Once released, this special object unloads every method call the client made to it into a special message that's stored in a queue supported by Microsoft Message Queue (MSMQ), a service that's independent of COM. When the laptop is finally connected to a network, MSMQ automatically locates the machine the real server is on and sends the message there. The COM+ environment on the target machine can be setup to listen for new messages and will grab this one when it comes in. COM+ then uses QC.Player to create and replay everything the original client did on the laptop.

And you get all of this functionality with little change to your client code and no change to your server.

Example 40   Writing and Using a COM+ Queued COM Server   **297**

Figure 10.8   Configure a COM+ Component to "Subscribe" to an Event Class

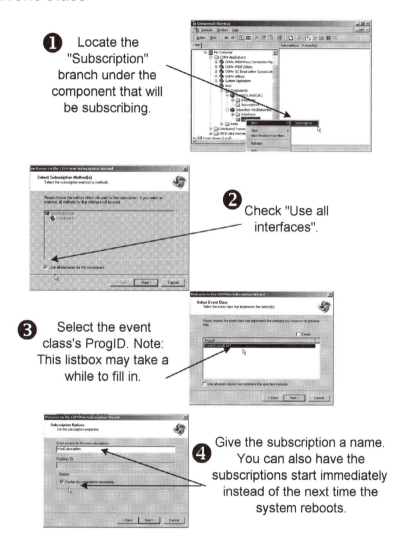

1 Locate the "Subscription" branch under the component that will be subscribing.

2 Check "Use all interfaces".

3 Select the event class's ProgID. Note: This listbox may take a while to fill in.

4 Give the subscription a name. You can also have the subscriptions start immediately instead of the next time the system reboots.

5 You can configure the subscriber to be a queued component by opening the subscription's properties and clicking here.

II

10

Steps

Configure the Queued Server

1. Within the Component Services, right-click on the application that contains the queued server (the real server the client wants to talk to). In the "Queuing" tab, checkmark both items. This causes COM+ to create the QC.ListenerHelper to monitor the MSMQ for any new queued messages.

2. You will also need to tell COM+ that this interface by opening its properties sheet and selecting "Queued" on the "Queuing" tab.

3. COM+'s QC.ListenerHelper will now monitor MSMQ for this application. To review, please see Figure 10.9.

Setup the Client

1. First, you need to register the dll server on the client as usual using OLE/COM Object Viewer, specifying the machine name where the real server is located.

2. Next you invoke QC.Recorder to do its job by using CoGetObject() instead of CoCreateInstance() or smart pointers. The syntax is as follows:

```
IWzd *pWzd;
HRESULT  hr=
        ::CoGetObject(L"queue:/new:Server.Wzd",
                           //ProgID of the real server

        0,
        __uuidof(IWzd),      //Interface ID of the real server
        (LPVOID*)&pWzd);     //pointer to real server
if (FAILED(hr))
{
    _com_error err(hr);
    AfxMessageBox(err.ErrorMessage());
    return;
}
IWzdPtr pPtr(pWzd);
pPtr->Method1(1234);
```

Example 40   Writing and Using a COM+ Queued COM Server   **299**

Figure 10.9   Enabling a COM+ Server to Receive Queued Input

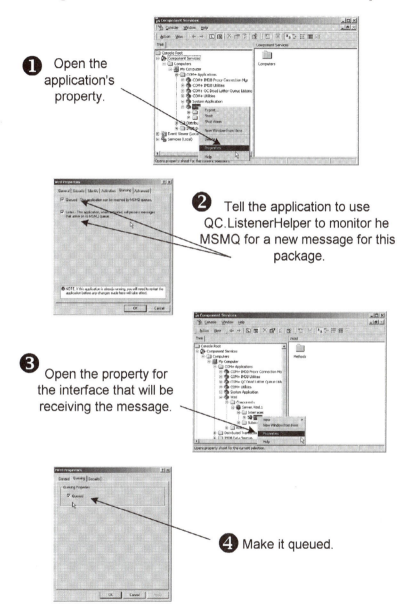

① Open the application's property.

② Tell the application to use QC.ListenerHelper to monitor he MSMQ for a new message for this package.

③ Open the property for the interface that will be receiving the message.

④ Make it queued.

II

10

3. COM+ gets the destination computer's name from how you registered the dll on the client system. However, you can also specify the computer name in the `CoGetObject()` call's binary string like so:

```
L"queue:ComputerName=mycomputer/new:Server.Wzd
```

Notes

- Please see Chapter 4 for more on queued components. You may feel a little leery about storing up your data as meaningless messages in MSMQ. However, COM+ makes a tremendous effort to get this data to the real server even making use of a dead-letter office. And using COM+ is a lot easier than the alternative — such as maintaining a mini-database on the laptop and what that entails.

Chapter 11

Accessing Database Objects

Eventually Microsoft hopes to replace the entire Win32 API with COM objects. Instead of creating a window by calling the global function `::CreateWindow()`, you would create a COM object which has a method called `CreateWindow()` among others that you would call. (Of course, to create the COM object you would still have to call the static function `::CoCreateInstance()`, but you've got to start somewhere.) One of the first API functions to be replaced by COM has been the ODBC API that allows universal access to data sources available to your application. The COM objects that were written to take over this functionality are collectively called OLE DB. And these objects were so versatile, so COM-like, and so basic that it was immediately realized that they were just too complicated to use by anyone other than their original programmers. So a new layer of COM classes was written to help simplify and encapsulate the functionality of OLE DB and these classes were collectively called Active Data Objects or ADO.

What you immediately find with ADO is that there's about a hundred ways to do anything. You can open a database, then create a command object to open a recordset object, or you can just open a command object and then open a recordset or you can just create a recordset object and specify the database and table right in its argument list. In this chapter, I give but one fairly comprehensive way to open a database and examine its records. And since ADO also supports late binding, I try to duplicate this comprehensive approach in Visual C++, Visual Basic, and Visual J++.

The examples in this chapter therefore include:

Example 41   Accessing a Database Using C++ and ADO   where we explore how to use ADO to access a database.

Example 42   Accessing a Database Using Visual Basic and ADO where we do the same for VB.

Example 43   Accessing a Database Using Visual J++ and ADO where we do the same for VJ++.

Example 41   Accessing a Database Using C++ and ADO

Objective

You would like to use the Active Data Object (ADO) COM server to access a database from your C++ application.

Strategy

Although there are dozens of ADO classes and methods, 10% of those classes and methods are used 90% of the time. We will therefore review just the basic everyday functionality you need — open the database, get a recordset and scroll through it, etc. In fact, if you have my previous book (*VC++ MFC Extensions by Example*, 1999, R&D Books), the example here is just the ADO version of the ODBC and DAO examples I show there.

Example 41   Accessing a Database Using C++ and ADO   **303**

Steps

Import ADO's Definitions

1. Import the type library for ADO into the source file in which you will be using ADO:

```
#import "c:\Program files\Common files\System\ADO\MSADO15.dll"
    rename("EOF", "_EOF")
using namespace ADODB;
```

Initialize your Client

1. Remember that ADO is a COM object like any other and you need to initialize COM for your application before you can use it. So call the following to initialize the COM for your main process _and_ call from any thread that will be using COM:

```
::CoInitializeEx(
NULL,                       //always NULL
COINIT_APARTMENTTHREADED    //see book about threading models
    );
```

Note: You also need to add _WIN32_DCOM to your project settings under "Preprocessor definitions" in order to get the prototype definition for ::CoInitializeEx() included in your compile.

Open and Close the Database

1. Start by creating the ADO COM objects for connecting and issuing commands to the data source:

```
_CommandPtr command(__uuidof(Command));
_ConnectionPtr connection(__uuidof(Connection));
```

2. To open a data source connection, use:

```
connection->PutConnectionTimeout(
            1000        //default is DEFAULT_DBCONNECTPROP
            );
connection->Open(
    L"WZDDB",    // connection string DSN (other ex:"MSOracle8")
    L"",         // user id
    L"",         // password
    adCmdUnspecified
    );
// initialize command object
    command->PutRefActiveConnection(connection);
```

3. To close the data source connection:

```
connection->Close();
```

Open and Access a Recordset for a Table

1. To get a recordset from the data source using SQL, use:

```
    _RecordsetPtr rs(__uuidof(Recordset));
// specify SQL command
    command->PutCommandText(L"SELECT * FROM CUSTOMERS");
    command->PutCommandType(adCmdText);
// what text in command specifies. also:
                    // adCmdStoredProc - stored procedure
                    // adCmdTable - table name
                    // adCmdFile - file name
                    // adCmdTableDirect -
                    // adCmdUnknown - unknown


// open record set
    rs->Open((_Command*)command,                     // source of
    record set--can be a SQL statement, etc.--see last param
            vtMissing,  // active connection object (optional)
            adOpenForwardOnly,// cursor type. also:
                // adOpenForwardOnly
                //    adOpenKeyset Opens a keyset-type cursor.
```

Example 41   Accessing a Database Using C++ and ADO   **305**

```
           //    adOpenDynamic Opens a dynamic-type cursor.
           //    adOpenStatic Opens a static type cursor.
        adLockOptimistic, // record set locking type. also:
        // adLockReadOnly -- read only
        // adLockPessimistic --lock records at the data source
                // when editing
        // adLockOptimistic -- lock records only
                // when you call the Update method.
        // adLockBatchOptimistic -- same as above,
                // used with batch mode
        adCmdUnspecified  // define source param—
                //here we can use a command object. also:
        // adCmdText indicates source is a SQL statement
        // adCmdTable indicates source is a table name.
        // adCmdStoredProc indicates source as a stored
                // procedure.
        // adCmdUnknown indicates source argument is not
                // known.
        );
```

2. To scroll through the recordset:

```
while (!rs->Get_EOF())
{
    _variant_t CustomerID=rs->GetCollect(L"CustomerID");
    _variant_t CompanyName=rs->GetCollect(L"CompanyName");
    _variant_t ContactName=rs->GetCollect(L"ContactName");
    rs->Move(1);            //# of records to move (can be negative
                            //  if adOpenDynamic in Open())
}
rs->MoveFirst();
```

3. To add a record to the recordset, use:

```
// will allow these depending on the locking mode the record
set was opened with
if (rs->Supports(adAddNew) && rs->Supports(adUpdate))
{
rs->AddNew();
```

```
rs->PutCollect(L"CustomerID",L"BONCO");
rs->PutCollect(L"CompanyName",L"Franks");
rs->PutCollect(L"ContactName",L"Runk");
rs->Update();
```

4. To modify a field in a row in a recordset:

```
rs->PutCollect(L"ContactName",L"Smith");
rs->Update();
```

5. To delete a record from the recordset:

```
rs->Delete(
adAffectCurrent  //can also be adAffectGroup
        // which deletes all rows in Filter object
    );
```

6. To close the recordset:

```
rs->Close();
```

Use a Store Procedure

1. To setup ADO with the parameters necessary to call a stored procedure:

```
// specify stored procedure
command->PutCommandText(L"Customers Query");
command->PutCommandType(adCmdStoredProc);

// in order and one-by-one, add the calling parameter(s)
required by the stored procedure
command->GetParameters()->Append(command->CreateParameter(
    _bstr_t(""),// a user assigned parameter name (optional)
    adInteger,  // data type. other common ones:
            //     adUnsignedInt
            //     adBSTR
            //     adArray|adBinary -- blob
        adParamInput,       // direction. also:
            //     adParamOutput
            //     adParamInputOutput
            //     adParamReturnValue
```

Example 41   Accessing a Database Using C++ and ADO   **307**

```
    5,      // database column size for this parameter
            //  required by SP (must be a long)
    _variant_t(1234L)        //if adParamInput or
      // adParamInputOutput, the input value as a variant
    ));
```

2. Execute the stored procedure:

```
// specify timeout (optional)
command->PutCommandTimeout(
    1000//default is DEFAULT_DBCONNECTPROP
    );
// execute stored procedure
command->Execute(
    NULL, // returned number of records affected
    NULL, // an array of variants that can be used with command
          //  (in lieu of appending above)
    adCmdStoredProc   // what text in command specifies. also:
            // adCmdText - SQL statement
            // adCmdTable - table name
            // adCmdFile - file name
            // adCmdTableDirect -
            // adCmdUnknown - unknown
        );
```

3. Retrieve any output parameters back from this stored procedure call if the parameter direction is adParamOutput, adParamInputOutput, or adParamReturnValue. In this example, we will retrieve from the zeroth parameter in the stored procedure call:

```
long vt=command->GetParameters()->GetItem(_variant_t(0L))
->GetValue().vt;
```

or retrieve as a variant

```
_variant_t vart=command->GetParameters()->GetItem(_variant_t(0L))
->GetValue().Detach();
```

II

11

Use a Store Procedure that Returns a Recordset

1. Specify the stored procedure's argument list marking the spot where the recordset argument will appear with "`resultset`":

```
command->PutCommandText(L"{ call Wzd.Package.SP(?,?,?,?,{resultset
    1024, my_out})}");
command->PutCommandType(adCmdText);
```

2. Add input parameters only, one for each "?" in the stored procedure argument list:

```
command->GetParameters()->Append(command->CreateParameter(
    _bstr_t(""),// a user assigned parameter name (optional)
    adInteger,        // data type. other common ones:
 :    :    :
```

3. The recordset is returned in an ADO `_RecordSet` class which you can scroll through and close as shown above or here:

```
rs->Open((_Command*)command, vtMissing, adOpenForwardOnly,
         adLockReadOnly, adCmdUnspecified);
rs->Close();
```

Use Transactions

If your COM server will be used in an MTS/COM+ application, please refer to Example 36 (page 276) for how to use transactions. However, outside of that framework, ADO itself supports transactions as seen in this section.

1. To start a transaction, use:

```
connection->BeginTrans();
```

2. To commit a transaction, use:

```
connection->CommitTrans();
```

3. Or to rollback a transaction, use:

```
connection->RollbackTrans();
```

Process ADO Errors

ADO's error handling is more sophisticated than most COM objects. If several errors caused an ADO method to fail, you can actually scroll through

Example 41   Accessing a Database Using C++ and ADO   **309**

each error, because ADO stores each and every error in a data collection. (Of course, each error will probably be enigmatic, still....)

1. To catch and scroll through ADO errors:

```
try
{
    ...using ADO COM objects...
}
catch(_com_error& e)
{
    ErrorsPtr pErrors=connection->GetErrors();
    if (pErrors->GetCount()==0)
    {
        AfxMessageBox(e.ErrorMessage());
    }
    else
    {
        for (int i=0;i<pErrors->GetCount();i++)
        {
            _bstr_t desc=pErrors->GetItem((long)i)
->GetDescription();
            AfxMessageBox(desc);
        }
    }
}
catch(...)
{
}
}
```

Notes

- This example depends on you configuring your database on your machine using ODBC32 in the Control Panel. You can also use the "Pro-vider=" keyword to cause ADO to bypass ODBC entirely. In the "Open a Database" section previously in step 2, you would use "Provider=xxx; Data source = xxx.mdb", where *xxx.mdb* is the actual mdb filename.

II

11

However, in order for this syntax to work, *xxx* must a valid `ProjID` of a COM object that can take the place of ODBC. Microsoft provides such a COM object for both Access databases and SQL Server databases. (For Access, it is: `Provider=Microsoft.JET.OLEDB.4.0; Data source = xxx.mdb.`) Please refer to your database vendor for information on any Provider COM objects they might supply.

• Let's talk about blobs for a second. For a binary blob, use an ADO data type of `adLongVarBinary`, use an Access database type of OLE Object, and use a SafeArray in your application to contain it. For a string blob, which can be as big as a binary blob but can only contain alphanumeric characters (no zeros), use an ADO data type of `adLongVarWChar`, an Access database type of Memo, and a BSTR data type in your application. And don't forget to destroy the SafeArray or BSTR when you're done with them.

• As mentioned, there are thousands of permutations of how to accomplish anything with ADO. Microsoft's ADO web site is loaded with a few of these: `http://www.microsoft.com/data/ado`.

CD Notes

• First, use ODBC to setup the dummy Access database that comes with this example. When executing the project, set a breakpoint in the `OnTest()` button handler and step through as ADO accesses that database.

Example 42   Accessing a Database Using Visual Basic and ADO

Objective

You would like to use the Active Data Object (ADO) COM interface to access a database from your Visual Basic application.

Strategy

Step-by-step we will almost duplicate everything we did in the last example using C++, only now with a Visual Basic spin. That means no need to worry about the fact that you're using a COM object.

Example 42   Accessing a Database Using Visual Basic and ADO   **311**

Steps

Define the ADO Classes

1. As with any COM server, use the "Project/References" menu commands to add ADO to your project. The server itself is called: Microsoft ActiveX Data Objects Library.

2. Define the ADO classes to be used:

```
Dim cn As New ADODB.Connection
Dim cmd As New ADODB.Command
Dim rs As New ADODB.Recordset
Dim prm As ADODB.Parameter
```

II

Open and Close the Database

1. To open a data source connection, use:

```
    cn.ConnectionTimeout = 1000
' the parameters below are: connection string, user id, password,
    cn.Open "WZDDB", "", "", adCmdUnspecified
' initialize command object
    Set cmd.ActiveConnection = cn
```

2. To close the data source connection:

```
    cn.Close
```

Access a Recordset

1. To get a recordset from the data source using SQL, use:

```
    rs.Open "SELECT * FROM CUSTOMERS", cn, adOpenForwardOnly,
    adLockPessimistic
```

2. To scroll through the recordset:

```
    rs.MoveFirst
    While (Not rs.EOF)
        CustomerID = rs("CustomerID")
```

11

```
            CompanyName = rs("CompanyName")
            ContactName = rs("ContactName")
        rs.MoveNext
    Wend
```

3. To add a record to the recordset, use:

```
    rs.AddNew
    rs("CustomerID") = "BONCO"
    rs("CompanyName") = "Franks"
    rs("ContactName") = "Runk"
    rs.Update
```

4. To modify a field in a row in a recordset; here we edit the current record:

```
    rs("CompanyName") = "Smith"
    rs.Update
```

5. To delete the current record from the recordset:

```
    rs.Delete
```

6. To close the recordset:

```
    rs.Close
```

Use a Stored Procedure

1. Specify the timeout (optional):

```
    cmd.CommandTimeout = 1000
```

2. Specify the stored procedure:

```
    cmd.CommandText = "Customers Query"
    cmd.CommandType = adCmdStoredProc
```

3. In order and one-by-one, add the calling parameter(s) required by the stored procedure. The parameters are the same as those in C++ ADO (please refer to the last example):

```
Set prm = cmd.CreateParameter("", adInteger, adParamInput, 5, 1234)
cmd.Parameters.Append prm
```

Example 42   Accessing a Database Using Visual Basic and ADO   **313**

4. Execute the stored procedure:

```
res = cmd.Execute(0, 0, adCmdStoredProc)
```

5. Retrieve an out parameter after the stored procedure is executed (if the parameter direction is adParamOutput, adParamInputOutput, or adParam-ReturnValue):

```
vt = rs(0) 'returns parameter 0
```

Use a Stored Procedure that Returns a Recordset

1. Specify the stored procedure to be called with a spot where the recordset will be returned marked with "resultset":

```
cmd.CommandText = "{ call Wzd.Package.SP(?,?,?,?,{resultset
1024, my_out})}"
cmd.CommandType = adCmdText
```

2. Add input parameters only, one for each "?" in the stored procedure call (see above):

```
Set prm = cmd.CreateParameter("", adInteger, adParamInput, 5, 1234)
cmd.Parameters.Append prm
```

3. Execute the stored procedure:

```
Set rs = cmd.Execute(0, 0, adCmdStoredProc)
```

4. Scroll through rs as with any recordset (see the "Access a Recordset" section previously).

5. Close the recordset:

```
rs.Close
```

Use Transactions

1. To start a transaction, use:

```
cn.BeginTrans
```

2. To commit a transaction, use:

```
cn.CommitTrans
```

II

11

3. To rollback a transaction, use:

```
cn.RollbackTrans
```

> **Note:** Sometime in the future, you will hopefully be able to write VB COM servers for MTS/COM+. At that point, you would also be able to use their transaction services instead of these.

Process ADO Errors

To catch and scroll through ADO errors:

1. At the top of the subroutine, use:

```
On Error GoTo AdoError
```

2. Then, at the bottom, use:

```
goto Done 'option—you could just let it drop through
AdoError:
        Dim errLoop As Error
        Dim strError As String
        Set Errs = cn.Errors
        For Each errLoop In Errs
                Debug.Print errLoop.SQLState
                Debug.Print errLoop.NativeError
                Debug.Print errLoop.Description
        Next
Done:
```

Notes

• Please see the notes under the last example.

CD Notes

• First, use the ODBC32 Control Panel Applet to setup the example's mdb file with a DSN of "WZDDB". Then, set a breakpoint at the top of the form routine and step through as it accesses the database.

Example 43   Accessing a Database Using Visual J++ and ADO   **315**

Example 43   Accessing a Database Using Visual J++ and ADO

Objective

You would like to use the Active Data Object (ADO) COM interface to access a database from your Visual J++ application.

Strategy

Step-by-step we will almost duplicate everything we did in the last example using C++ only now with a Visual J++ spin. That means no need to worry about the fact that you're using a COM object.

Steps

Define the ADO Classes

1. Import the ADO class definitions into any source file that will be using ADO:

```
import com.ms.com.*;
import com.ms.wfc.data.*;
```

2. Create the ADO COM objects for connecting and issuing commands to the data source:

```
com.ms.wfc.data.Command command=new com.ms.wfc.data.Command();
com.ms.wfc.data.Connection connection=
                new com.ms.wfc.data.Connection();
```

Open and Close the Database

1. To open a data source connection, use:

```
connection.setConnectionTimeout(
        1000            //default is DEFAULT_DBCONNECTPROP
        );
connection.open(
        "WZDDB",// connection string DSN
```

```
                        //(other ex:"MSOracle8")
        "",                 // user id
        ""                  // password
    );
// initialize command object
      command.setActiveConnection(connection);
```

2. To close the data source connection:

```
    connection.close();
```

Access a Recordset

1. To get a recordset from the data source using SQL, use:

```
        com.ms.wfc.data.Recordset rs =new Recordset();
// specify SQL command
      command.setCommandText("SELECT * FROM CUSTOMERS");
      command.setCommandType(AdoEnums.CommandType.TEXT);
          // what text in command specifies. also:
          // adCmdStoredProc - stored procedure
          // adCmdTable - table name
          // adCmdFile - file name
          // adCmdTableDirect -
          // adCmdUnknown - unknown
// open record set
      rs.open(command,   // source of record set--can be
                     // a SQL statement, etc.--see last param
        null,    // active connection object (optional)
        AdoEnums.CursorType.FORWARDONLY,// cursor type. also:
            // adOpenForwardOnly
            //   adOpenKeyset Opens a keyset-type cursor.
            //   adOpenDynamic Opens a dynamic-type cursor.
            //   adOpenStatic Opens a static type cursor.
        AdoEnums.LockType.OPTIMISTIC,
                // record set locking type. also:
            // adLockReadOnly -- read only
            // adLockPessimistic --lock records at the
                        // data source when editing
```

Example 43   Accessing a Database Using Visual J++ and ADO   **317**

```
                    // adLockOptimistic -- lock records only when
                         //you call the Update method.
                // adLockBatchOptimistic -- same as above, used
                         //with batch mode
          AdoEnums.CommandType.UNSPECIFIED
                    // define source param--here we can use a
                    // command object. also:
                // adCmdText indicates source is a SQL statement
                // adCmdTable indicates source is a table name.
                // adCmdStoredProc indicates source as a stored
                         //procedure.
                // adCmdUnknown indicates source argument is not
                         //known.
                );
```

II

2. To scroll through the recordset:

```
while (!rs.getEOF())
{
    String CustomerID=rs.getField("CustomerID").getString();
    String CompanyName=rs.getField("CompanyName").getString();
    String ContactName=rs.getField("ContactName").getString();
    rs.moveNext();
}
rs.moveFirst();
```

3. To add a record to the recordset, use:

```
rs.addNew();
rs.getField("CustomerID").setString("BONCO");
rs.getField("CompanyName").setString("Franks");
rs.getField("ContactName").setString("Runk");
rs.update();
```

11

4. To modify a field in the current row in a recordset:

```
rs.getField("ContactName").setString("Smith");
rs.update();
```

5. To delete the current record from the recordset:

```
rs.delete(AdoEnums.Affect.CURRENT);
```

6. To close the recordset:

```
rs.close();
```

Use a Stored Procedure

1. Specify a timeout (optional):

```
command.setCommandTimeout(
1000        //default is DEFAULT_DBCONNECTPROP
    );
```

2. Specify the stored procedure:

```
command.setCommandText("Customers Query");
command.setCommandType(AdoEnums.CommandType.STOREDPROC);
```

3. In order and one-by-one, add the calling parameter(s) required by the stored procedure:

```
command.getParameters().append(command.createParameter(
        "",     // a user assigned parameter name (optional)
        AdoEnums.DataType.INTEGER,          // data type.
        AdoEnums.ParameterDirection.INPUT,// direction.
        5,          // database column size for this parameter
                // required by SP (must be a long)
        new Variant(1234)              //if adParamInput or
        // adParamInputOutput, the input value as a variant
        ));
```

4. Execute the stored procedure:

```
command.execute(null, // returned number of records affected
    AdoEnums.CommandType.STOREDPROC // what text in command
                                    // specifies.
    );
```

Example 43   Accessing a Database Using Visual J++ and ADO   **319**

5. After the stored procedure executes, retrieve an out parameter if parameter direction is adParamOutput, adParamInputOutput, or adParamReturn-Value:

```
// retrieve a long from the zeroth parameter
long vt=command.getParameters().getItem(0).getValue().getvt();
//retrieve the 0 parameter
// retrieve a variant the zeroth parameter
Variant vart=command.getParameters().getItem(0).getValue();
```

Use a Stored Procedure that Returns a Recordset

1. Specify the stored procedure with a spot where the recordset will be returned marked with "resultset":

```
command.setCommandText("{ call
Wzd.Package.SP(?,?,?,?,{resultset 1024, my_out})}");
command.setCommandType(AdoEnums.CommandType.TEXT);
```

2. Add input parameters only, one for each "?" in stored procedure call above:

```
command.getParameters().append(command.createParameter(
        "",      // a user assigned parameter name (optional)
        AdoEnums.DataType.INTEGER,           // data type.
        AdoEnums.ParameterDirection.INPUT,// direction.
        5,            // database column size for this parameter
                      // required by SP (must be a long)
        new Variant(1234) //if adParamInput or
                      // adParamInputOutput, the input
                      // value as a variant
    ));
```

3. Open, scroll through and close recordset as shown above:

```
rs.open(command,null,AdoEnums.CursorType.FORWARDONLY,AdoEnums.
LockType.OPTIMISTIC,AdoEnums.CommandType.UNSPECIFIED);
rs.close();
```

Use Transactions

1. To open a transaction, use:

```
connection.beginTrans();
```

2. To commit a transaction, use:

```
connection.commitTrans();
```

3. To rollback a transaction:

```
connection.rollbackTrans();
```

Note: Sometime in the future, you will hopefully be able to write VJ++ COM servers for MTS/COM+. At that point, you would also be able to use their transaction services instead of these.

Process ADO Errors

1. To catch and scroll through ADO errors:

```
try
{
    ...using ADO COM objects...
}
catch(AdoException ae)
{
    for (int i = 0; i< connection.getErrors().getCount(); i++)
    {
        com.ms.wfc.data.Error errItem =
connection.getErrors().getItem(i);
        System.out.println(errItem.getDescription());
        System.out.println("\nPress <Enter> to continue...");
    }
}
catch (Exception ae)
{
}
```

Example 43   Accessing a Database Using Visual J++ and ADO   **321**

Notes

- Please see the notes for the first example in this chapter.

CD Notes

- First use the ODBC32 Control Panel Applet to setup the example's mdb file with a DSN of "WZDDB". Then, set a breakpoint at the top of the form routine and step through as it accesses the database.

II

Chapter 12

Potpourri

This is the "everything else" chapter and COM has a lot of those. Because your work can be potentially a part of some other application, at what point would you check to see if that application has a license to run your server? There's also the whole subject of reporting errors back to your client, especially when there might be several layers of COM servers between where the error occurred and where the client made the first call. And finally, there's the arcane subject of the "free-threaded marshaller" — the most talked about subject you'll never use.

The examples in this chapter include:

Example 44   Licensing Your COM Object Using MFC where   your COM server refuses to create an instance of itself until it checks to make sure the client has a license.

Example 45   Licensing Your COM Object Using ATL where   we   do the same for a COM server created with ATL.

Example 46   Processing COM Errors where we look at several schemes for reporting errors back to a client.

Example 47   Turning off Marshalling for a "Both" COM Object Using MFC   where we prevent COM from ever using marshalling on our object.

Example 48   Turning off Marshalling for a "Both" COM Object Using ATL   where we do the same for COM servers written in ATL.

Example 44   Licensing Your COM Object Using MFC

Objective

You would like to prevent your COM server from being used by a client that doesn't have a license to do so.

Strategy

The MFC class `COleObjectFactory` makes it very easy to add licensing logic to your COM server. You just override its `VerifyUserLicense()` member function and return false if the client doesn't have a license. The major problem, however, is being able to derive a class from `COleObjectFactory`. The name is declared statically inside of an MFC macro forcing you to rewrite that macro. Fortunately, it's a small macro.

Steps

Write the COM Server

1. Create your MFC COM class as usual using the ClassWizard.
2. To the top of the generated `.h` file of this COM class, add your own derivation of the MFC `COleObjectFactory` class as follows:

```
class CWzdOleObjectFactory : public COleObjectFactory
{
public:
    CWzdOleObjectFactory( REFCLSID clsid, CRuntimeClass*
    pRuntimeClass,
        BOOL bMultiInstance, LPCTSTR lpszProgID ) :
```

Example 44   Licensing Your COM Object Using MFC   **325**

```
        COleObjectFactory(clsid,pRuntimeClass,bMultiInstance,
    lpszProgID)
            {};
}
```

3. Within the braces of this class, you will be overriding the `VerifyUserLicense()` method like so:

```
    virtual BOOL VerifyUserLicense()
    {
        return AfxVerifyLicFile(AfxGetInstanceHandle(),
            "licence.lic",
            L"1234567890",
                -1);
    }
```

Where "`license.lic`" is a license file that exists in the same directory as the COM server's DLL or EXE file and contains a license key as a string of numbers. And `L"1234567890"` is what should be in that license key file.

4. Just below this class, add two new macro definitions that are based on two MFC OLE macros. These macros with this new `CWzdOleObjectFactory` class:

```
#define DECLARE_OLECREATE_WZD(class_name) \
public: \
    static AFX_DATA CWzdOleObjectFactory factory; \
    static AFX_DATA const GUID guid; \

#define IMPLEMENT_OLECREATE_WZD(class_name, external_name, l, w1, \
    w2, b1, b2, b3, b4, b5, b6, b7, b8) \
    AFX_DATADEF CWzdOleObjectFactory
    class_name::factory(class_name::guid, \
        RUNTIME_CLASS(class_name), FALSE, _T(external_name)); \
    AFX_COMDAT const AFX_DATADEF GUID class_name::guid = \
        { l, w1, w2, { b1, b2, b3, b4, b5, b6, b7, b8 } };
```

5. Substitute these macros for the current macros. In this same .h file, find the `DECLARE_OLECREATE` macro and rename it to `DECLARE_OLECREATE_WZD`. In the .cpp class file, replace the `IMPLEMENT_OLECREATE` macro with `IMPLEMENT_OLECREATE_WZD`.

6. For a complete listing of this `.h` file, please refer to the end of this example.

Notes

- You might think this strategy a bit lame considering someone could just copy the license file with the COM DLL or EXE file. So you might make it more secure by adding a lot more logic to `VerifyUserLicense()`. Perhaps make the license file expire or base the license on some facet of the machine it's installed on.

CD Notes

- Build the project and put a breakpoint in `VerifyUserLicense()` and watch as the license is checked before the object is created.

Listings — COM Class `.H` File

```
#if
    !defined(AFX_WZDSRV_H__4487D433_A6FF_11D3_A398_00C04F570E2C__INCLUDED_)
#define
    AFX_WZDSRV_H__4487D433_A6FF_11D3_A398_00C04F570E2C__INCLUDED_

#if _MSC_VER > 1000
#pragma once
#endif // _MSC_VER > 1000
// WzdSrv.h : header file
//
#include "afxctl.h"
#include "..\IServer\IWzd.h"


///////////////////////////////////////////////////////////////////
    /////////
// Implement our own derivation of COleObjectFactory class factory

class CWzdOleObjectFactory : public COleObjectFactory
{
public:
    CWzdOleObjectFactory( REFCLSID clsid, CRuntimeClass*
    pRuntimeClass,
```

Example 44   Licensing Your COM Object Using MFC   **327**

```
        BOOL bMultiInstance, LPCTSTR lpszProgID ) :
            COleObjectFactory(clsid,pRuntimeClass,bMultiInstance,
    lpszProgID)
            {};


    virtual BOOL VerifyUserLicense()
    {
        return
    AfxVerifyLicFile(AfxGetInstanceHandle(),"licence.lic",
    L"1234567890",-1);
    }
};



#define DECLARE_OLECREATE_WZD(class_name) \
public: \
    static AFX_DATA CWzdOleObjectFactory factory; \
    static AFX_DATA const GUID guid; \


#define IMPLEMENT_OLECREATE_WZD(class_name, external_name, l, w1, \
    w2, b1, b2, b3, b4, b5, b6, b7, b8) \
    AFX_DATADEF CWzdOleObjectFactory
    class_name::factory(class_name::guid, \
        RUNTIME_CLASS(class_name), FALSE, _T(external_name)); \
    AFX_COMDAT const AFX_DATADEF GUID class_name::guid = \
        { l, w1, w2, { b1, b2, b3, b4, b5, b6, b7, b8 } }; \


/////////////////////////////////////////////////////////////////
    ////////
// CWzdSrv command target

class CWzdSrv : public CCmdTarget
{
    DECLARE_DYNCREATE(CWzdSrv)

    CWzdSrv();          // protected constructor used by dynamic creation
```

```
// Attributes
public:

// Operations
public:

// Overrides
    // ClassWizard generated virtual function overrides
    //{{AFX_VIRTUAL(CWzdSrv)
    public:
    virtual void OnFinalRelease();
    //}}AFX_VIRTUAL
    BOOL VerifyUserLicense();

// Implementation
protected:
    virtual ~CWzdSrv();

    // Generated message map functions
    //{{AFX_MSG(CWzdSrv)
        // NOTE - the ClassWizard will add and remove member
    functions here.
    //}}AFX_MSG

    DECLARE_MESSAGE_MAP()
    DECLARE_OLECREATE_WZD(CWzdSrv)

    // Generated OLE dispatch map functions
    //{{AFX_DISPATCH(CWzdSrv)
        // NOTE - the ClassWizard will add and remove member
    functions here.
    //}}AFX_DISPATCH
    DECLARE_DISPATCH_MAP()
    DECLARE_INTERFACE_MAP()
    BEGIN_INTERFACE_PART(WzdClass, IWzd)
        STDMETHOD_(HRESULT,Method1)(long, long *);
        STDMETHOD_(HRESULT,Method2)(long, unsigned long);
```

Example 45   Licensing Your COM Object Using ATL   **329**

```
        END_INTERFACE_PART(WzdClass)

};


////////////////////////////////////////////////////////////////////
    /////////

//{{AFX_INSERT_LOCATION}}
// Microsoft Visual C++ will insert additional declarations
    immediately before the previous line.

#endif //
    !defined(AFX_WZDSRV_H__4487D433_A6FF_11D3_A398_00C04F570E2C__IN
    CLUDED_)
```

II

Example 45   Licensing Your COM Object Using ATL

Objective

You would like to prevent your COM server from being used by a client that doesn't have a license to do so.

Strategy

ATL makes it a lot easier to add licensing support to your COM server using one of its ubiquitous macros `DECLARE_CLASSFACTORY2()`. However, you still need to write your own licensing class.

Steps

Write the COM Server

1. Create the ATL server and object as usual using the ATL Object Wizard.

12

2. Manually add a new class to the top of the created object's .h file that looks like this:

```
class CWzdLicense
{
protected:
    static BOOL VerifyLicenseKey(BSTR bstr)
    {

        // compare bstr with embedded license info
        return TRUE; // valid license

    }

    static BOOL GetLicenseKey(DWORD dwReserved, BSTR* pBstr)
    {

      *pBstr = L"licence";// embedded license info
      return TRUE;

    }

    static BOOL IsLicenseValid()
    {

        // open license file and compare with license key info
      embedded in this class
        return TRUE; //TRUE==licensed

    }
};
```

3. In the IsLicenseValid() method above, add logic that opens a file on disk that checks to see if this object is running on a licensed system.

4. Fill the VerifyLicenseKey() method with a comparison of the input BSTR string and a valid license.

5. Return the license in the GetLicenseKey() method. Note that as a protected method, this is not available to the client, which would obviously be fairly self-defeating, but is available to your object.

6. Also add a macro to this .h file that references this class:

```
DECLARE_CLASSFACTORY2(CWzdLicense)
```

7. Please refer to the end of this example for a complete listing of this .h file.

Example 45   Licensing Your COM Object Using ATL   **331**

Notes

- Please see the notes under the last example.

CD Notes

- Build the project and put a breakpoint in `VerifyUserLicense()` and watch as the license is checked before the object is created.

Listings — COM Class .H File

```
// Wzd.h : Declaration of the CWzd

#ifndef __WZD_H_
#define __WZD_H_


#include "resource.h"        // main symbols



//////////////////////////////////////////////////////////////////////
    ////////
// CWzdLicense-- a class that checks on licensing
class CWzdLicense
{
protected:
   static BOOL VerifyLicenseKey(BSTR bstr)
   {
       // compare bstr with embedded license info
       return TRUE; // valid license
   }

   static BOOL GetLicenseKey(DWORD dwReserved, BSTR* pBstr)
   {
     *pBstr = L"licence";// embedded license info
     return TRUE;
   }

   static BOOL IsLicenseValid()
```

II

12

```
    {
        // open license file and compare with license key info
    embedded in this class
        return TRUE; //TRUE==licensed
    }
};


//////////////////////////////////////////////////////////////////
    ////////
// CWzd
class ATL_NO_VTABLE CWzd :
    public CComObjectRootEx<CComSingleThreadModel>,
    public CComCoClass<CWzd, &CLSID_Wzd>,
    public IDispatchImpl<IWzd, &IID_IWzd, &LIBID_SERVERLib>
{
public:
    CWzd()
    {
    }


DECLARE_CLASSFACTORY2(CWzdLicense)


DECLARE_REGISTRY_RESOURCEID(IDR_WZD)


DECLARE_PROTECT_FINAL_CONSTRUCT()


BEGIN_COM_MAP(CWzd)
    COM_INTERFACE_ENTRY(IWzd)
    COM_INTERFACE_ENTRY(IDispatch)
END_COM_MAP()

// IWzd
public:
    STDMETHOD(Method1)(long lArg);
};
#endif //__WZD_H_
```

Example 46   Processing COM Errors   **333**

Example 46   Processing COM Errors

Objective

You would like to use ATL's error handling techniques to report errors back to your client.

Strategy

This example shows a little bit of everything. We catch the COM error class _com_error, which is thrown by smart pointer classes and ADO. We write our own user-defined COM error codes using a COM macro. And we use two COM API calls, SetErrorInfo() and GetErrorInfo(), to pass a IErrorInfo COM object from our server to our client for much more robust error reporting.

Steps

Catch Thrown Errors

1. To catch COM errors thrown by smart pointers and other COM classes that use the _com_error class, use:

```
try
{
    retval=pPtr->Method1(1234);
}
catch (_com_error &err)
{
    AfxMessageBox(err.ErrorMessage());
}
```

2. When using a smart pointer to access a COM server, the server methods you call are actually smart pointer methods which check for a COM failure. When a COM failure is found (FAILED()), this method wrapper then throws an exception using the _com_error class. To avoid this check and potential throw, you can access the unadulterated server method directly by calling it directly with:

```
hr=pPtr->raw_Method1(1234,&retval);
```

where Method1() is the original method name.

3. Or you could prevent any method from being wrapped by using the raw_interfaces_only keyword when you use the #import directive at the top of the listing.

Create Your Own COM Error Codes

1. You can create your own COM error codes, usually in your own errors.h file like so:

```
#if !defined errors_h
#define errors_h

const HRESULT WZD_ERROR1 = MAKE_HRESULT(
                    1,      // 1=failure, 0=warning
                    FACILITY_ITF,
// COM errors (can also be FACILITY_WINDOWS for window errors, etc.)
                    0x0400                 // user defined 0x400 and above
                    );
const HRESULT WZD_ERROR2 = MAKE_HRESULT(1, FACILITY_ITF, 0x0401);
const HRESULT WZD_ERROR3 = MAKE_HRESULT(1, FACILITY_ITF, 0x0402);
#endif
```

If you make the first argument to this macro a zero (0), and return it to your client, it will pass the FAILED() test, but still inform the client of something. A classic use of this is to tell a client when the server has reached the end of a recordset or a file.

Return More Error Information to Your Client

The COM API comes with two API calls which allow you to create and send an IErrorInfo object from your COM server to the client. This IErrorInfo object can contain a lot more information on the cause of an error other than an error number. You can pass an actual error message, the location of the error, and even the location in a help file for the client to look up. To use these API calls:

Example 46   Processing COM Errors   **335**

1. When an error occurs, create your own error message like so:

```
ICreateErrorInfo *cei;
HRESULT hr = CreateErrorInfo(&cei);
IErrorInfo *eip;
hr = cei->QueryInterface(IID_IErrorInfo, (LPVOID*) &eip);
if (!FAILED(hr))
{
    cei->SetDescription(L"Something bad happened.");
            // the error in human form
    cei->SetSource(_bstr_t(__FILE__));
            // where it happened
    cei->SetHelpContext(23);
        // plug into help file to summon a
    cei->SetHelpFile(L"help.hlp");
        // particular passage
    SetErrorInfo(0, eip);
    eip->Release();
}
cei->Release();
```

2. Then return a failure to let the client know that something bad happened so it knows to look for this message:

```
    return WZD_ERROR3;
```

3. In the client, you can retrieve this error like so:

```
IErrorInfoPtr eip(pPtr);
hr = GetErrorInfo(0, &eip);
if (!FAILED(hr))
{
    DWORD dwHelpContext;
    BSTR bstrDescription,bstrSource,bstrHelpFile;
    eip->GetDescription(&bstrDescription);
        // error in english
    eip->GetSource(&bstrSource);
        // where did the error occur
    eip->GetHelpContext(&dwHelpContext);
```

II

12

```
            // plug into help file to summon a
       eip->GetHelpFile(&bstrHelpFile);
       // particular passage

}
```

Even though you've returned an error, how does a client know that you have also created an `IErrorInfo` object for it? Unless you're writing both server and client, you need to provide some way for the client to ask the COM server if it has an `IErrorInfo` object for it. The official way to do this is to include an `ISupportErrorInfo` COM class in your COM server. This class has but one method: `InterfaceSupportsErrorInfo()` that returns a COM failure if your COM server doesn't. You can automatically add this class to your COM server in ATL:

1. When creating the COM class using the ATL Object Wizard, specify "Support `ISupportErrorInfo`". This simply adds the `ISupportErrorInfo` COM class to your server with an implementation that says, "Yes". It does nothing else for you.

2. In the client, when you receive an error, create a new safe pointer from the object's pointer for `ISupportErrorInfo` like so:

```
ISupportErrorInfoPtr sei(pPtr);
```

where `pPtr` is the smart pointer for the COM object that had the error. This is the first chance for the client to determine whether the COM server has created an `IErrorInfo` — if the `QueryInterface()` that this smart pointer performs fails, the COM server doesn't support it.

3. Then, just before using `GetErrorInfo()` to retrieve the `IErrorInfo` object, check `ISupportErrorInfo`'s one method:

```
hr=sei->InterfaceSupportsErrorInfo(__uuidof(IWzd));
if (hr==S_OK)
{
// retrieve IErrorInfo
:     :    :
}
```

where `IWzd` is the interface of the COM server your using. Even if the COM server does create the `IErrorInfo` object, it might not have done it for this error. The `InterfaceSupportsErrorInfo()` checks to see if the server used `SetErrorInfo()` and if not, a failure is returned. And `GetErroInfo()` can only be called once.

Notes

- Some clients such as a web browser will automatically display the error message in your `IErrorInfo` object.

- Using `GetErrorInfo()` and `SetErrorInfo()` only works for the current thread. That means your client won't be able to grab an `IErrorInfo` object created in an out-of-process EXE or DLL, and if it was created by a COM object in another thread of your application. So rather than use `GetErrorInfo()` and `SetErrorInfo()`, you might consider passing the `IErrorInfo` object along yourself in your method argument list. Or you might send it to some error handler service, such as an NT service, that will log your errors. Or you might even consider replacing `IErrorInfo` with your own custom error information object.

CD Notes

- When executing the project on the accompanying CD, notice that the error message created in the server is passed to the client.

Example 47   Turning off Marshalling for a "Both" COM Object Using MFC

Objective

You would like to prevent COM from ever marshalling your COM server's methods.

Strategy

As mentioned in Chapter 3, COM can be used to automatically make your COM servers thread safe. One way that COM does this is by marshalling your COM server's methods, even though the server is a DLL in the same process space as its client. Because marshalling is the exact same process that goes on when COM sends data to a COM server that's in another application or another country, this thread safety comes at the price of some additional processing overhead. So if you've already made your COM server thread safe (e.g., by adding critical sections to your code), you don't need to incur this performance hit. Why would COM ever marshal your server when it doesn't need it? If you configure it as a "Both" threading

II

12

model. "Both", as you recall, causes a COM server to be used as both an apartment-threaded and free-threaded object depending on the object that creates it. But because a "Both" object must be written as thread safe already to be able to be used as a free-threaded object, adding marshalling to it in the other situation is a waste of CPU cycles. Fortunately, you can get around this situation by simply aggregating your COM server with what's called the "free-threaded marshaller", which is a COM server provided by the system that takes over when COM is about to add marshalling to your object. In fact, COM thinks it is still marshalling your server; it's just that the free-threaded marshaller is directly calling your methods instead of doing all that other stuff.

Note: Using the free-threaded marshaller with a COM server that will be out-of-process makes no sense — your COM server's methods must be marshalled for the original purpose of marshalling, just to get your data from the client to the server.

Steps

Use the Free-Threaded Marshaller

1. Follow Example 16 (page 173) for MFC aggregation — except the COM class you'll be aggregating will be a system COM class.

2. At the point where you would use `::CoCreateInstance()` to create the object to aggregate, use the following instead:

```
::CoCreateFreeThreadedMarshaler(GetControllingUnknown,&pUnknown);
```

3. That's it — this object will only be marshalled if it's out-of-process.

Notes

- As neat as this one is, you'll probably never use it if you do things right.

Example 48   Turning off Marshalling for a "Both" COM Object Using ATL

Objective

You would like to prevent COM from ever marshalling your COM server's methods.

Strategy

Refer to the last example and Chapter 3 for more on the "free-threaded marshaller". Adding one to your ATL COM server is just a setting you need to make when you use the ATL Object Wizard.

Steps

Use the Free-Threaded Marshaller

1. Create the ATL server as usual with the ATL AppWizard.
2. Create the ATL COM class as usual with the ATL Object Wizard, except on the second tab make sure to specify "Free Threaded Marshaller".
3. That's it — this object will only be marshalled if it's out-of-process.

Notes

- It's been a pleasure. I'll see you all at the beach.

II

12

Appendix A

COM Tables

Table A.1 Sending Simple Message Types

What to send	IDL type	C++ type	VB type	VJ++ type
Unsigned 8 bits	char	char	Byte	char
Unsigned 8 bits	boolean	bool	(1)	(1)
Unsigned 8 bits	byte	BYTE	(1)	(1)
Unsigned 16 bits	unsigned short	unsigned short	(1)	(1)
Signed 16 bits	short	short	Integer	short
Unsigned 32 bits	unsigned long	unsigned long	(1)	(1)
Signed 32 bits	long	long	Long	int
Unsigned 64 bits	unsigned hyper	unsigned hyper	(1)	(1)
Signed 64 bits	hyper	hyper	(1)	(1)
32-bit floating point	float	float	Single	float

What to send	IDL type	C++ type	VB type	VJ++ type
64-bit floating point	double	double	Double	double
Enumerator	Enum(2)	Enum(2)	Enum(2)	Enum(2)
Currency (8 bytes)	CY	CY	CY	CY
Date (double)	DATE	DATE	DATE	DATE

NOTE 1: Although IDL might allow you to define a certain argument type, VB and VJ++ also depend on the COM dll to know how to send that type. In this case, OLE32.DLL doesn't know what to do and you can't sepcify a proxy/stub dll for the late binding that VB and VJ++ supports so this type isn't supported. What happens if COM realizes it needs a proxy/stub DLL but can't find it or load it? No, there's no error. COM just tries to communicate what it can and leaves the rest behind. Believe it or not.

NOTE 2: Before you use an enumerator in your argument list, you must first define it at the top of the IDL file before any other declarations like so:

```
// enumerator (above any interface definitions)
typedef enum {Monday=2, Tuesday, Wednesday, Thursday, Friday} workday;
```

Then use it in your argument list like any other type:

```
method( [in] workday enumArg );
```

The enumerator's definition is automatically included any time you include this COM class definition.

Table A.2 Sending Simple Arrays

Type of Array	IDL declaration	C++ declarations
Fixed size	`long aArg[25]`	`long aArg[25]`
Variable sized	`[in,size_is(lSize)] long aArg[][25]`	`long lSize, long **aArg`
Variable sized/limited transmisson	`[in,first_is(lFirst), last_is(lLast), size_is(lSize)] long *aArg[][25]`	`long lFirst, long lLast, long lSize, long **aArg`
Variable sized/limited transmisson	`[in,first_is(lFirst), length_is(lLength), size_is(lSize)] long *aArg[][25]`	`long lFirst, long lLength, long lSize, long **aArg`

Table A.3 Sending Visual Basic Types

What to send/receive from VB	IDL type	VC++ type	VJ++ type
text strings	BSTR	BSTR or _bstr_t class	String class
arrays	SAFEARRAY(BYTE)	SAFEARRAY *	Safearray class
untyped arguments	VARIANT	VARIANT or _variant_t class	Variant class

Table A.4 In-Process Thread Safety Using COINIT_APARTMENTTHREADED

This Entity...	Creating an object with this threading model...	Creates the new object in this thread...	And method calls to the object are....
The main process	Single/Apartment/Both	Main process	Direct
	Free	New	Synchronized
A thread	Single	Main Process	Synchronized
	Apartment/Both	Same	Direct
	Free	New	Synchronized
Single object	Single/Apartment/Both	Same	Direct
	Free	New	Synchronized
Apartment or Both object running in main process	Single/Apartment/Both	Same	Direct
	Free	New	Synchronized
Apartment or Both object running in a thread	Single	Main Process	Synchronized
	Apartment/Both	Same	Direct
	Free	New	Synchronized

A

This Entity...	Creating an object with this threading model...	Creates the new object in this thread...	And method calls to the object are....
Free object	Single	Main Process	Synchronized
	Apartment	New	Synchronized
	Free/Both	Same	Direct

Table A.5 In-Process Thread Safety Using `COINIT_MULTITHREADED`

This Entity...	Creating an object with this threading model...	Creates the new object in this thread...	And method calls to the object are....
The main process or thread	Single/Apartment	Communal	Synchronized
	Free/Both	Same	Direct
Single object	Single/Apartment/Both	Same	Direct
	Free	New	Synchronized
Apartment object running in main process	Single/Apartment/Both	Same	Direct
	Free	New	Synchronized
Apartment object running in a thread	Single	Main Process	Synchronized
	Apartment/Both	Same	Direct
	Free	New	Synchronized
Free/Both object	Single/ Apartment	Communal	Synchronized
	Free/Both	Same	Direct

Table A.6 Out-of-Process Thread Safety Using `COINIT_APARTMENTTHREADED`

This Entity...	Creating an object with this threading model...	Creates the new object in this thread...	And method calls to the object are....
Any External Client	Any	Communal	Synchronized
Any COM object (Single,etc.)	Any	Same	Direct

Table A.7 Out-of-Process Thread Safety Using `COINIT_MULTITHREADED`

This Entity...	Creating an object with this threading model...	Creates the new object in this thread...	And method calls to the object are....
An external Client	Any	New	Synchronized
Any COM object (Single,etc.)	Any	New Communal	Synchronized

A

Appendix B

COM Troubleshooting Guide

Table B.1

Symptom	Possible Problem	Solutions
HRESULT returned is 0x800401f0	The client isn't calling CoInitInstance(), which must be called from an application *and* any thread created by the application before it can use a COM function (Co...) except for the CoTaskMemAlloc() family of calls.	Call CoInitInstance() from all of your threads, John.
HRESULT returned is 0x80040154	The COM server's DLL or EXE hasn't been registered on your system. Even when the DLL or EXE you want to access is running on another machine, you must still register it on your machine and then use OLEVIEW to redirect your system to the other system. When first working with COM, this will be the root of most of your problems.	If the server is a DLL, use "regsvr32 xxx.dll" to register the DLL, where *xxx* is the name of the DLL. If an EXE, usually just executing the EXE once is enough to register it. Whenever possible, let the system automatically register your COM servers for you.
HRESULT returned is 0x80004002	The COM class you're trying to create doesn't exist in this DLL or EXE or it's being identified with the wrong Interface ID.	If you're not using smart pointers, double check the Interface ID against the value in the IDL file of the COM server. If you're using smart pointers, make sure the type library you're importing isn't out-of-date, or make a point of importing from the dll or exe file directly.
You're debugging and when you try to step into a method call you get this error: "The value of ESP was not properly saved across a function call."	The number of arguments in the method that your client is calling is different then what the server is expecting.	Make sure the type library you're importing isn't out-of-date and that the dll or exe file that is registered on your system is the one your client application is importing.

Symptom	Possible Problem	Solutions
Warning MIDL2039: interface does not conform to [oleautomation] attribute	One of the argument types you specified isn't supported by the COM DLL. You will get this error if you specified the IDL is for a dual interface.	Your client can still use an object created with this interface if it uses early binding, however if it uses it with late binding, the offending argument will not be properly transmitted from client to server and back — but even then, if the server is a DLL, you're using in-process and you don't have to worry.
All of the data in an argument, such as an array or a structure, isn't getting to the server or vice versa.	You are passing an argument type that is not in the COM DLL's repertoire of argument types and therefore requires you to supply a proxy/stub dll.	You have possibly not created this dll, or forgotten to register it on both the client machine *and* the server machine.
You set a breakpoint in a COM server, but it disappears when you run the application.	When your application is using a COM server, it doesn't actually link with a COM DLL until runtime and it never links with a COM EXE. You can tell the debugger to allow you to set breakpoints in a COM DLL by going into your application's settings.	Open up the "Debug" tab and select "Additional DLLs" from the "Category" combo box. Then browse and select the dll file(s) you want to be able to set breakpoints for. With an EXE file, all you can do is bring it up in its own Studio, set breakpoints in it and run it as a separate project. If the EXE is already running, COM will attach to it rather then create a new EXE. However, make sure to start the COM EXE before you start the client application, otherwise COM will start a new EXE.

B

Symptom	Possible Problem	Solutions
When you run an application, the debugger keeps telling you that such-and-such a dll is missing.	If you add a dll to your project settings using the steps in the last solution, but that dll gets moved or deleted, the debugger will complain. Non-COM dll's that are linked to your application, but are missing from your execution path, will also complain.	Either return the dll to where the settings thinks it is or delete it from your project settings. For non-COM dll's, just add them to your execution path (current directory or a subdirectory specified by your environments "PATH=" statement).
When compiling your ATL server, you get this compiler error: "error C2259: 'CComObject<class XXX>' : cannot instantiate abstract class due to following members:"	The argument lists of your methods in the IDL, .H and .CPP file *must match exactly*.	Put all three in an editor and compare or keep eliminating methods until the error goes away.
When working with ATL, the ClassView won't allow you to add a new method/property.	Make sure the IDL, ,H and .CPP files aren't read-only. When you use a source control application like Source-Safe, your files are kept read-only until they are checked out for edits.	Just check all three files out.
When linking your ATL server for a "Release" configuration, you get this error: "LIBCMT.lib(crt0.obj) : error LNK2001: unresolved external symbol _main"	Your ATL server is using a function in the C runtime library that requires initialization, which is done in the _main function. Unfortunately, when an ATL server is compiled for minimum size, the _main function is left out of the link.	Open your project settings and remove the ATL_MIN_CRT for the Release configurations of your build.

Symptom	Possible Problem	Solutions
When stepping into a method call of an out-of-process server, the HRESULT returned is C0000005 or you get an "Access violation" or the system hangs.	A pointer argument is pointing to garbage causing the proxy/stub or COM DLL to try to copy garbage to the server. You won't get this error with in-process calls because COM doesn't bring the proxy/stub into play.	Make sure all pointer arguments are getting a valid address or NULL.
	The COM server's EXE hasn't been registered on the remote system.	Register on remote system by running it once.
	The COM server's DLL hasn't been registered on the remote system with a DLL surrogate such as DLL-HOST, MTS, or COM+.	Register it.
	When creating a COM object on another system, the same account and password that you logged into your current machine must also exist on the server machine. That means if you logged in using a user id of "joe" and password of "1234", the same account and password must exist on the machine that has the COM DLL or EXE server.	Add the correct account and password to the server machine.

Your particular application may present you with unique problems. One of the worst feelings in the world is remembering you ran across a problem before but forget how you fixed it, or you wrote it down but forget where you put it. Here's a place to write it down — just don't lose the book.

B

Table B.2 Your Notes

Symptoms	What made it work

Index

Symbols

`#import` directive 16–17, 110–113, 121, 123, 334

`[ptr]` 43

`[ref]` 43

`[unique]` 43

`_bstr_t` 43–44, 256, 258–259, 306, 308–309, 335, 343

`_variant_t` 44, 258–259, 305, 307, 343

A

Access Permissions 59

`Activate()` 89, 279–280

Activation Security 58

Active Data Objects (ADO) 278–279, 281, 301–303, 306, 308–312, 314–315, 320, 333
See Chapter 11

Active Template Library (ATL) 24, 85, 191–192, 199

ActiveX 4–5, 24, 26, 28, 51, 93, 103–104, 123–127, 132–133, 142, 166, 184–190, 233–236, 240, 285, 311

Events 51

Wizard 26

`AddRef()` 15, 17, 20, 26, 149, 155, 164–165, 183

ADO web site 310

`AfxConnectionAdvise()` 47, 163, 165–166

`AfxVerifyLicFile()` 60

Aggregation 56–57, 173, 226–227, 229
See also Encapsulation

Apartments 71

AppID 10–11, 27

AppWizard 25–26, 85, 136, 147–148, 157–161, 193–194, 198, 200, 203–204, 277, 339

ATL Object Wizard 48–49, 57, 89, 193–194, 207–208, 212, 220–222, 227, 277, 329, 336, 339

Attribute Programming 98

Automatic Thread Safety 64

Automation 26, 104, 141, 143, 158–159, 188, 195, 213, 259

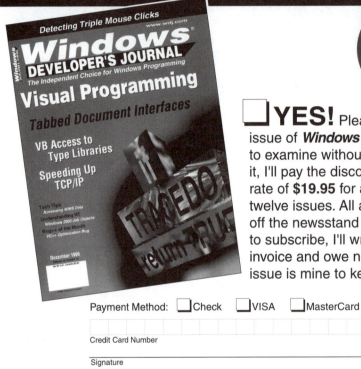

RD3003   **$49.95**

Visual C++ MFC Programming by Example

by John E. Swanke

Learn by example how to extend MFC to create more sophisticated and powerful applications. First you learn the in's and out's of messaging, which is the single most important concept to understanding how MFC works. Then the author presents 85 user interface examples, each fully annotated and ready to insert into applications. "Any half dozen of the 85 examples is likely to save you more than a couple of hours of research." Ron Burk *Windows Developer's Journal.* CD-ROM and MFC Quick Reference Guide included, 593pp, ISBN 0-87930-544-4

VC++ MFC Extensions by Example

by John E. Swanke

Extend MFC to create more sophisticated and powerful applications. You get 67 examples — each fully annotated and ready to insert into applications. This book features a menu of advanced techniques across the entire range of Windows functions that complement the author's earlier title, *Visual C++ MFC Programming by Example*. The CD contains working projects in Visual C++ V5.0 & V6.0 and the author's own *SampleWizard* utility that facilitates adding these examples into users' applications. CD and MFC Quick Reference Guide included, 643pp, ISBN 0-87930-588-6

RD3250   **$49.95**

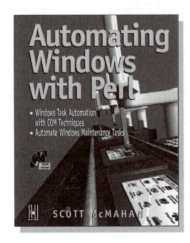
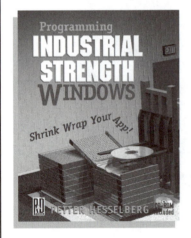

RD2985   **$29.95**

Debugging Visual C++ Windows
by Keith Bugg

Understand and control the Bug cycle! This detailed reference provides tutorial-based examples and a conceptual model for preventing and eliminating bugs during the design cycle that includes suggestions on identification, prevention, and correction for each of the four types of bugs: compile-time errors, run-time errors, logic and design errors, and machine errors. You will learn how compiler tools such as TRACE, Spy, and Stress work and get a critical review of commercial debuggers (including BoundsChecker and Code Wizard). Disk included, 206pp, ISBN 0-87930-545-2

Supercharge MFC
by Jeffrey Galbraith

Supercharge your MFC development of enhanced Windows User Interface (UI) controls with pseudo-multiple inheritance. Because message processing is the core of Windows programs, the author has created a sophisticated, yet generic, message handler that can be used with any window — eliminating the need for complicated MFC message maps. This subclassing method works directly with MFC objects. If MFC should do the work, it lets the message pass through; if you want your own class extensions to do the work, then the C++ wrappers handle the messages themselves. CD-ROM included, 520pp, ISBN 0-87930-569-X

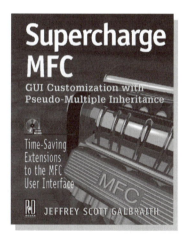

RD3056   **$49.95**

What's on the CD-ROM?

COM Programming by Example is accompanied by the companion CD-ROM which contains a working Visual C++, Visual Basic, or Visual J++ v6.0 project for every example in the book. If you want to find the project for an example, just locate its example number among the subdirectories on the CD.

About the SampleWizard™

Also on the CD is the SampleWizard utility which can help you add the examples in the book directly to your applications. This utility guides you through a catalog of examples which, if selected, details the instructions and code necessary for including the example in your project. You will also be given the opportunity to substitute the example's project name ("Wzd") with your own.

The SampleWizard can be found in the \SWD directory on the CD. It makes use of the \Wizard subdirectory found in each example on the CD and contains all of the particulars for that example. Simply execute SW.EXE. The rest should be intuitive.

It is particularly useful as a user tool in the Developer Studio. Make sure to configure the directory as the current project directory [$(WkspDir)] and your requested example will be copied directly into your project directory.

For additional information on using **SampleWizard**™, see the ReadMe file on the CD. Or contact the author at http://home.earthlink.net/~jeswanke.
